T0231278

BIOCATALYSIS AND AGRICULTURAL BIOTECHNOLOGY

BIOCATALYSIS AND AGRICULTURAL BIOTECHNOLOGY

Fundamentals, Advances, and Practices for a Greener Future

Anjali Priyadarshini, PhD
Prerna Pandey, PhD

Apple Academic Press Inc.   Apple Academic Press Inc.
3333 Mistwell Crescent   9 Spinnaker Way
Oakville, ON L6L 0A2   Waretown, NJ 08758
Canada   USA

© 2019 by Apple Academic Press, Inc.

First issued in paperback 2021

Exclusive worldwide distribution by CRC Press, a member of Taylor & Francis Group
No claim to original U.S. Government works

ISBN 13: 978-1-77463-168-3 (pbk)
ISBN 13: 978-1-77188-689-5 (hbk)

Library and Archives Canada Cataloguing in Publication

Priyadarshini, Anjali, author
Biocatalysis and agricultural biotechnology : fundamentals, advances, and practices for a greener future / Anjali Priyadarshini, PhD, Prerna Pandey, PhD.

Includes bibliographical references and index.
Issued in print and electronic formats.
ISBN 978-1-77188-689-5 (hardcover).--ISBN 978-1-351-16744-4 (PDF)

1. Agricultural biotechnology. 2. Biocatalysis. 3. Enzymes--Biotechnology.
I. Pandey, Prerna, author II. Title.

| S494.5.B563P75 2018 | 664'.024 | C2018-903385-1 | C2018-903386-X |

CIP data on file with US Library of Congress

Apple Academic Press also publishes its books in a variety of electronic formats. Some content that appears in print may not be available in electronic format. For information about Apple Academic Press products, visit our website at **www.appleacademicpress.com** and the CRC Press website at **www.crcpress.com**

CONTENTS

ABOUT THE AUTHORS

Anjali Priyadarshini, PhD
Assistant Professor, SRM University, Haryana, India;
74-B, Ayodhya Enclave, Sector 13 Rohini, New Delhi–85, India,
E-mail: anjalipriyadarshini1@gmail.com; anjali0419@yahoo.co.in

Anjali Priyadarshini, PhD, is a Council of Scientific & Industrial Research (CSIR), Government of India awardee and holds a doctorate in biomedical sciences. Her field of research and interest includes biotechnology and nanotechnology. Dr. Priyadarshini has published papers in peer-reviewed journals in the biomedical field. She currently works as assistant professor at SRM University, Haryana, India.

Prerna Pandey, PhD
Biotechnologist and Writer, Maharashtra, India;
B-1403, Jasper, Hiranandani Estate, Thane, Maharashtra,
India–400607, Tel.: +919619423223,
E-mail: prernapandey@gmail.com, prernapandey@hotmail.com

Prerna Pandey, PhD, a biotechnologist with several years of wet lab research experience. She has worked at the International Center for Genetic Engineering and Biotechnology, New Delhi, India. Her field included isolation and molecular characterization of Geminiviruses, genome sequencing, gene annotation, and gene silencing using the RNA interference technology. She has also worked at Transasia Biomedicals and Advance Enzyme Technologies as a scientist. Dr. Pandey has published papers in peer-reviewed journals in the field and has submitted a number of annotated Geminiviral genome sequences to the GenBank, including two novel ones. She has also completed her editing and proofreading courses from Society for Editors and Proofreaders (SfEP) and now works as a freelance scientific writer and editor.

LIST OF ABBREVIATIONS

AAT	acyltransferase
ABA	abscisic acid
ACVS	ACV synthetase
AD	Alzheimer's disease
ADDS	advanced drug delivery systems
ALDC	acetolactate decarboxylases
AMF	arbuscular mycorrhiza fungi
AP	alkaline phosphatase
BR	brassinosteroids
BRC	British Retail Consortium
CAT	catalase
CCR	cinammoyl-CoA reductase
CGF	The Consumer Goods Forum
CK	cytokinins
CMS	cytoplasmic male sterility
CT	conventional tillage
DAOC	deacetoxycephalosporin C
DLD	drosomycin-like defensin
DMSO	dimethyl sulfoxide
DNA	deoxyribonucleic acid
EFGF	electric field gradient focusing
EFSA	European Food Safety Authority
EG	ethylene glycol
EIQ	Environmental Impact Quotient
EMA	effective microbial agent
EPC	European Patent Convention
ER	endoplasmic reticulum
FAD	flavin adenine dinucleotide
FAO	Food and Agriculture Organization
GA	Golgi apparatus
GBS	genotyping-by-sequencing

GFSI	Global Food Safety Initiative
GHG	greenhouse gas
GM	genetically modified
GMS	genetic male sterility
GOD	glucose oxidase
GR	golden rice
GST	glutathione S-transferase
HDV	hepatitis delta virus
HRP	horseradish peroxidase
IAA	indole acetic acid
IAFP	International Association for Food Protection
IARC	International Agriculture Research Centers
IET	initial evaluation trail
IL	ionic liquid
IPNS	isopenicillin N synthase
LAT	L-lysine aminotransferase
LEA	late embryogenesis abundant
MS	microspheres
MSU	Michigan State University
NDV	Newcastle disease virus
NT	no-till
ODM	oligonucleotide-directed mutagenesis
OECD	Organization for Economic Co-operation and Development
PA	penicillin acylase
PBR	Plant Breeder's Rights
PCR	polymerase chain reaction
PEG	polyethylene glycol
PEG	polyethyleneglycol
PGPR	plant growth promoting rhizobacteria
PLGA	poly D, L-lactide-co-glycolide
PNK	polynucleotide kinase
PRSV	papaya ringspot-virus
PVPA	Plant Variety Protection Act
PVS	plant vitrification solution
RNA	ribonucleic acid

RT	reverse transcriptase
SA	salicylic acid
SARP	streptomyces antibiotic regulatory protein
SOD	superoxide dismutase
TERI	Tata Energy Research Institute
TF	transcription factors
USDA	United States Department of Agriculture
UV	ultraviolet
UVT	uniform variety trail
VS	Varkud satellite
WTO	World Trade Organization
ZFN	zinc-finger nucleases

PREFACE

This new volume, *Biocatalysis and Agricultural Biotechnology: Fundamentals, Advances, and Practices for a Greener Future*, looks at the application of a variety of technologies, both fundamental and advanced, that are being used for crop improvement, metabolic engineering, and the development of transgenic plants.

The book is divided into two sections. Part 1, on biocatalysis and agriculture biotechnology, covers the fundamentals and the latest advances in the field of biocatalysis, an interdisciplinary subject that includes aspects of both organic and inorganic chemistry. This section covers a range of topics from enzymology to different classes of enzymes and their applications in their native or immobilized state (as whole cells in aqueous as well as nonconventional media) to an in-depth description of catalytic mechanisms. Techniques such as "white biotechnology" and the fact that biocatalysis is one of the main prerequisites for a sustainable development are also discussed.

Part 2 of the book covers agricultural biotechnology and novel agricultural practices. Conventional and advanced practices are discussed in detail, including the scope and history of agricultural biotechnology, crop improvement practices, plant tissue culture, genetic modification for crop improvement, and production of transgenic crops, as well as regulation and patenting of plant products, etc. The authors also discuss the requirements of global food safety and their importance in today's world.

The volume provides a holistic view that makes it a valuable source of information for researchers of agriculture and biotechnology, along with agricultural engineers and environmental enthusiasts.

.

INTRODUCTION

A very warm greeting to our readers! During our years as students and as researchers, we often searched for books that would give us the comprehensive view about the topics on agriculture biotechnology, since this was not taught very elaborately in all the life science institutions. This book looks at the application of a variety of technologies – fundamental as well as the advanced – greatly being used for crop improvement, metabolic engineering and development of transgenic plants.

The science of agriculture is amongst the oldest and intensely studied by mankind. Human intervention has led to manipulation of plant gene structure to the use of plants for bioenergy. A sound knowledge of enzymology as well as the various biosynthetic pathway are required to further utilize microbes as sources to provide the desired products industrially for our utility. The chapters in this book thus give an overview of all these aspects and provide an updated review of the major plant biotechnology procedures and techniques, their impact on novel agricultural development, and crop plant improvement. We also discuss the use of "white biotechnology" and "metabolic engineering" as prerequisites for a sustainable development. The importance of patenting of plant products, world food safety, and the role of several imminent organizations have also been discussed.

We have tried to provide a holistic view and make the book as a valuable source of information not only to researchers of agriculture and biotechnology but also meets the course requirements of students in agriculture biotechnology, genetics, biology, biotechnology, and plant science; agricultural engineers, environmental biologists, environmental engineers, and environmentalists.

The book provides insight into several aspects of biocatalysis and agriculture biotechnology in brief as each topic in itself is an ocean of information. Nevertheless, each of the topics have been presented and elucidated in concise terms with relevant research. The book serves to ignite the minds of students and academicians to pursue research or just serve as reading material.

—*Anjali Priyadarshini, PhD*
Prerna Pandey, PhD

PART I

BIOCATALYSIS AND AGRICULTURAL BIOTECHNOLOGY

CHAPTER 1

ENZYMES

ANJALI PRIYADARSHINI and PRERNA PANDEY

CONTENTS

1.1 INTRODUCTION

Enzymes are biological molecules synthesized in a living cell that are responsible for catalyzing all important chemical interconversions that occurs in a cell. They are biological catalysts that can greatly accelerate both the rate and specificity of metabolic reactions, ranging from the digestion of food to the synthesis of DNA. Apart from its activity in the biological system, the enzymes are also capable of functioning (i.e., conduct catalysis) in vitro too. They can be produced in large amounts by microorganisms for industrial applications, and they are easily bio-degradable; thus, they do not add to the existing load of other non-biodegradable things posing a threat to the ecosystems present on the Earth.

Enzyme technology broadly involves:

   i.   production;
   ii. isolation;
   iii. purification;
   iv. modification of enzymes;
   v.   generation in soluble or immobilized form; and
   vi. production of more efficient and useful enzymes using recombi-nant DNA technology and protein engineering.

This can used for the ultimate benefit of humankind. The commercial production and use of enzymes is a major part of the biotechnology industry.

1.2   HISTORY OF ENZYMES

Extraction of rennet from calf's stomach by Danish chemist Christian Hansen in 1874 ushered the mankind into an era of modern enzyme tech-nology. This preparation goes into the history as probably the first enzyme preparation for industrial purposes. Usage of enzymes by humans can be traced back, when unknowingly for brewing, baking or curding enzymes were used in the form of microorganisms.

Historically, enzymatic processes involved in fermentation processes were the focus of numerous studies in the 19th century. An important dis-covery was the isolation of the enzyme complex from malt by Payen and Persoz in 1833, which converts gelatinized starch into sugars, primarily into maltose, and was termed "diastase." Development progressed par-ticularly in the field of fermentation where the achievements by Schwann, Liebig, Pasteur, and Kuhne were of great importance. Liebig claimed that fermentation occurred from chemical processes and that alive yeast cell is not required for the fermentation process, even if the yeast is not viable its enzyme bring about the reaction. Pasteur, on the other hand, argued that fermentation did not occur unless viable organisms were present.

The dispute was finally settled in 1897, after the death of both adversaries, when the Buchner brothers demonstrated the conversion of glucose into ethanol and carbon dioxide by cell-free yeast extract and proved that yeast cells as such are not needed for such conversion, but the enzymes.

Year	Scientist	Contribution
1676	Antonie van Leewenhoek	Discovered microorganism with the aid of self made microsope
1858	Louis Pasteur	Isolated an optical isomer of tartaric acid using fermentation
1876	Kuhne	Gave the name enzyme
1893	Wilhelm Ostwald	Coined the term catalysis
1894	Emil Hermann Fischer	Proposed lock and key mechanism for enzyme action
1897	Eduard Buchner	Demonstated alcoholic fermentation by cell free extract
1907	Otto Rohm	Patented enzymatic treatment of leather with pancreatic extract and ammonium
1913	Leonard Michaelis and Maud Menton	Explained enzyme kinetics with equation
1926	James B. Sumner	Crystallization of Ureases enzyme from jack beans
1950	Pehr Edman	Developed polypeptide sequencing method
1951	Frederik Sanger	Elucidated amino acid sequence of insulin beta chain
1953	Francis H. C. Crick and J. D. Watson	Postulated double helix structure of DNA
1963	Stanford Moore and William Stein	Elucidation of amino acid sequence of lysozyme and ribonuclease
1960s	Werner Aber and Hamilton D. Smith	Discovery of restriction enzymes
1973	Stanley N. Cohen and Herbert Boyer	Performed gene experiments with restriction endonucleases

Year	Scientist	Contribution
1975	Cesar Milstein and Georges Köhler	Monoclonal antibody production
1985	Michael Smith	Changed amino acid sequence by site directed mutagenesis
1985	Kary B. Mullis	Invented PCR (polymerase chain reaction) technique
2001	Nature publication	Publication of sequencing result of human genome project

1.3   APPLICATIONS OF ENZYMES

Enzymes have a wide range of applications that include their use in food production, food processing, and preservation (pectinolytic enzymes have long been used to increase the yield and clarity of fruit juices), washing powders (lipase and protease), textile manufacture, leather industry, paper industry, medical applications (ELISA), improvement of environment and scientific research, and in bioelectronics where integration of redox enzymes with an electrode support and formation of an electrical contact between the biocatalyst and the electrode aids many function such in the development of biosensors, for example, glucose biosensor using glucose oxidase enzyme to monitor blood glucose level.

1.3.1   TEXTILE

In olden times, textiles were treated with acid, alkali, or oxidizing agents or soaked in water for several days to breakdown the starch without knowing the role of microorganisms in this process. This practice was difficult to control and sometimes also led to damage or discoloration of the material. Crude enzyme extracts in the form of malt extract, or later, in the form of pancreas extract, were first used to carry out designing, which was followed by enzymes from other sources.

There are a large number of microorganisms that produce a variety of enzymes [1, 2] helpful in carrying out many processes. The Textile industry, particularly the chemical processing sector, always has a major share in the global pollution. Enzymes play a key role in such alternative processes. The use of enzymes in textiles started as long as a century ago. Bacterial amylase derived from *Bacillus subtilis* was used for desizing for the first time by Boidin and Effront in 1917.

1.3.2 LEATHER BATING

Traditional method of leather bating in practice was to use the excrement of dogs and pigeons. This practice was changed by a German chemist and industrial Otto Rohm who proposed the theory that that these excrements had their effect because they contained residual amounts of the animal's digestive enzymes; thus, the extracts obtained from animals would suffice for such operation. This was proved to be correct when the bating process was done with great efficiency using pancreatic extracts that contained digestive enzymes, but later experiments showed that it was not the animals' enzymes that were active, but rather the enzymes of bacteria growing in the intestinal tract.

1.3.3 DETERGENT ENZYMES

Breakthrough in detergents was made in 1959, when Dr. Jaag, a chemist, developed a new product called Bio 40 that contained a bacterial protease instead of trypsin. Currently, these enzymes are manufactured commercially in large quantities through fermentation by common soil bacteria *Bacillus subtilis* or *Bacillus licheniformis*. This was made possible in the last two decades by the rapid advances in enzymology and fermentation technology. Although numerous other microorganisms produce proteases and amylases, the types secreted by the above strains have the advantage that they work best at the warm alkaline conditions prevailing in washing liquids. They also must not lose their activity in an environment that contains a multitude of potentially inhibitory chemicals routinely formulated into laundry detergents, such as surface active agents, magnesium or calcium ions, builders (sodium tripolyphosphate), perfumes, and other additives [3].

1.3.4   ENZYME TO CONVERT SUGARS FROM STARCH

Initially, fungal amylase was used in the manufacture of specific types of syrup, i.e., those containing a range of sugars, which could not be produced by conventional acid hydrolysis. This practice was changed in the 1960s when an enzyme glucoamylase was launched for the first time; this enzyme was capable of completely breaking down starch into glucose. Later, heat-stable alpha amylase development led to further improvement in this process.

A large number of cellulose-, starch-, and sugar-containing plants can be processed to produce sugars and alcohols, such as sugarcane, sweet sorghum, and nipa palm, which are the candidates for the high yield production of alcohol fuel. Likewise, the starch-containing crops such as cassava, sweet potatoes, yams, taro, and tannia are good candidates, but require an additional step to breakdown starch to sugar, the major part of biomass containing cellulose and which, therefore, needs special treatment before it can be used to produce glucose and alcohols

Years of research in biochemistry and biotechnology have boosted knowledge of enzymes for industries as well as research and led to the development of new techniques to modify and discover new application of enzymes in medicine, research, and industries. Thus, it has become the need of the day to device efficient methods for enzyme extraction as well as production for commercial purpose.

1.4   COMMERCIAL PRODUCTION OF ENZYMES

Presently, the majority (80%) of enzymes procured for commercial use are from microbial sources. Enzymes have also been obtained from animal and plant sources. Tables 1.1 and 1.2 list the enzymes with their sources and applications.

1.4.1   LIMITATIONS OF ENZYME EXTRACTION FROM ANIMAL AND PLANT SOURCE

The first and foremost limitation is the variation in their distribution and their limited quantity. This also involves high cost processes such as isolation, purification, and sterilization of the enzyme. In extraction of

TABLE 1.1 Commercial Enzyme Production from Plant Source and Their Application

Enzyme	Source(s)	Application(s)
β-amylase	Barley, soy bean	Baking, preparation of maltose syrup
Esterase	Wheat	Ester hydrolysis
Papain	Papaya	Meat tenderizer, tanning, baking
Peroxidase	Horse radish	Diagnostic
Urease	Jack bean	Diagnostic
Ficin	Fig	Meat tenderizer

TABLE 1.2 Commercial Enzymes from Animal Sources and Their Applications

Enzyme	Source	Application
Amylase, esterase	Lamb, calf	Digestive aid
Pepsin, trypsin	Bovine	Cheese preparation
Human urine	Urokinase	Dissolution of blood clot
Lysozyme	Hen egg	Cell wall breakage in bacteria
Lipase, renin, phospholipase, phytase	Porcine	Cheese preparation

industrial enzymes from animals such as bovine sources, there is always a risk of contamination with bovine spongiform encephalopathy (BSE is a prion disease caused by ingestion of abnormal proteins). Thus, due to these limitations of plant and animal source, microbial cells have become a choice source of enzyme for various purposes.

1.4.2 APPLICATIONS OF MICROBIAL ENZYMES

Microbial enzymes have found usage in food, pharmaceutical, textile, paper, leather, and various other industries. Most of the industrially important microbial enzymes are hydrolases, which catalyze the hydrolysis of natural organic compounds.

1.4.3 ENZYMES FROM MICROBIAL SOURCES

Microorganisms form the most significant and convenient sources of commercial enzyme production. They can be made to produce abundant quantities of enzymes under suitable growth conditions, by using inexpensive media. A selected list of enzymes, microbial sources, and the applications are given in Table 1.3.

1.5   PRODUCTION OF ENZYME USING MICROBIAL CELLS

Production of enzyme from microbial cells occurs via the processes listed below (Figure 1.1).

1. Selection of organisms depending upon the enzyme to be produced.
2. Formulation of medium according to the requirement of microorganism of choice.
3. Production process or fermentation (which may be of different types such as submerged or surface).
4. Recovery and purification of enzymes, which is also known as downstream processing.

1.5.1   SELECTION OF ORGANISM

The amount of enzyme produced by the microorganism remains the main criteria for its selection. The process becomes more appealing if the microbes produce and secrete the enzyme in the culture media, which makes it cost effective as this negates the use of techniques to isolate the cells and then extract the enzyme. Sometimes, the strains selected

TABLE 1.3   Industrially Important Enzymes from Microorganism and Their Application

Enzyme	Source	Application
α-amylase	*Aspergillus niger, A. oryzae, Baccilus subtilis, B. licheniforms*	Production of beer and alcohol, preparation of glucose syrups, digestive aid
Cellulase	*A. niger, Tricoderma koningi*	Alcohol and glucose production
Amyloglucosidase	*A.niger, Rhizopus niveus*	Starch hydrolysis
Invertase	*Saccharomyces cerevisiae*	Sucrose inversion, preparation of artificial honey, confectionaries

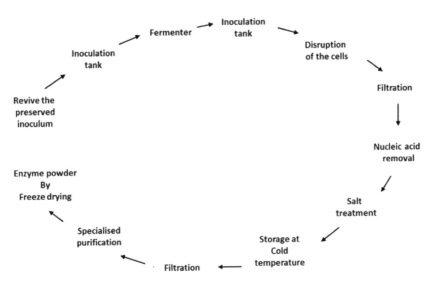

FIGURE 1.1 An outline of the production of enzymes by microorganisms. An inoculum containing the microorganism with the desired enzyme is used for production. After the product is formed, the microbial cells are disrupted and debris are removed by filtration, followed by subsequent removal of unwanted things and purification of the desired product. The final step is freeze drying and export.

for enzyme production need to be improved. Because of technological advancements, their genes can be altered with the use of mutagens (chemicals or UV rays) and desired changes could be brought about. Once the changes have been done and verified, the improved strain is used and provided to the manufactures.

1.5.2 FORMULATION OF FERMENTATION MEDIUM FOR ENZYME PRODUCTION

The pointer to be kept in mind here is the nutritional requirement of the microbial cell for the production of enzyme of interest. The nutrient that best suited for enzyme production (e.g., the substrate for enzyme) should be provided/ supplemented as all enzymes are not expressed constitutively, some are inducible. For the production of certain enzymes that are known as the inducible ones, the substrate has to be provided, and this can be done by supplementing it in the culture medium. Prior to this, the medium should contain all the nutrients for the robust growth and multiplication of microbial cells in question, and the raw material should be readily available, cost effective and safe to be used, such as starch hydrolysate, molasses, corn steep liquor, yeast extract, whey, and soy bean meal, apart from cereals (wheat) and pulses (peanut). In addition to the nutrition provided by the culture medium, its acidity or alkalinity has a major role to play in microbial growth, because microbial cells have an optimum pH range in which their growth and enzyme production is adequate. Thus, the maintenance of pH of the culture medium really becomes important in terms of production. Once the medium is prepared, it is sterilized using moist heat method (autoclaving) or filtration if it contains heat-labile substances.

1.5.3   PRODUCTION OR THE FERMENTATION PROCESS

Industrial production of enzymes is mostly carried out by submerged liquid conditions or by solid-substrate fermentation where the yields have more and less chances of infection. The production of enzymes is mostly carried out by batch fermentation and to a lesser extent by the continuous process. The sterile condition of the bioreactor system must be maintained throughout the fermentation process, which can vary from 2–7 days in most cases. The enzyme(s) have to be recovered and purified. Either surface or submerged culture methods currently may be employed for most microbial enzyme production. Usually, different cultures must be used for maximum enzyme yield by the two methods, although there are exceptions to this rule. There are advantages and disadvantages of each method, some of which are shown in Table 1.4.

1.5.4   DOWNSTREAM PROCESSING, RECOVERY, AND PURIFICATION OF ENZYMES

The downstream processing of the product varies according to the requirement and mode of production. The enzyme preparation might serve its purpose either in crude form or may be required in highly purified form. The desired enzyme produced may be excreted into the culture medium

TABLE 1.4 Comparison between Surface and Submerged Fermentation

Surface process	Submerged process
Labour intensive	Labour extensive
Less contamination prone	Contamination a major problem
Minimum control needed	Needs careful control of the process
Recovery involves extraction with aqueous solution, filtration or centrifugation sometimes evaporation and precipitation too needed	Recovery involves filtration, centrifugation, evaporation and precipitation
	Contamination a major problem
	Needs careful control of the process
	Recovery involves filtration, centrifugation, evaporation and precipitation

(extracellular enzymes) or may be present within the cells (intracellular enzymes), thus again dictating the process to be used. Recovery of an extracellular enzyme that is present in the fermentation broth is a simple process as compared to that for an intracellular enzyme. The broth needs to be filtered or centrifuged to remove the cellular part and then further purification done if needed. For intracellular enzymes, special techniques are needed for cell disruption for which sonication, high pressure application, glass beads, or lysozyme treatment are effective methods. The technique applied depends on its application as well as on the fact that it is extracellular or intracellular and this should result in a minimum loss of enzymatic activity.

1.5.5 REMOVAL OF NUCLEIC ACIDS

If the enzyme is produced intracellularly, the cells need to be disrupted for extraction. This results in the release of nucleic acid in the medium, which interferes with the recovery and purification of enzymes. To remove nucleic acids, polyamines are added for their precipitation.

1.5.6 ENZYME PRECIPITATION AND CONCENTRATION

Enzymes are present in the broth and needs to be concentrated using polyamines or polyethylene glycol. For this, enzymes can be precipitated using salts (ammonium sulfate) and organic solvents (isopropanol, ethanol, and acetone), which could then be dissolved into minimum quantity of the appropriate medium.

Chromatography is also a very advantageous technique in terms of degree of purification obtained. The chromatographic techniques for separation and purification of enzymes employed are:

i.   ion-exchange chromatography;
ii. size exclusion chromatography;
iii. affinity chromatography;
iv. hydrophobic interaction chromatography;
v.   dye ligand chromatography.

1.5.7 DRYING AND PACKING

The final step in the downstream processing is the concentration of the product. Drying with the help of film evaporators or use of freeze dryers (lyophilizers) is a very effective method. This method should not interfere with the stability of the enzyme. Stability of the dried product could further be ensured by keeping the dried enzyme in ammonium sulfate suspensions.

1.6 REGULATION OF MICROBIAL ENZYME PRODUCTION

The enzymes used in food or medical treatment should be totally free from toxic materials and harmful microorganisms, should not cause allergic reactions, must be of high grade purity, and must meet the required specifications by the regulatory bodies. As discussed earlier, some enzymes are produced constitutively by the cell and some needs to be induced. The induction can be done by using either the substrate or the product or any intermediate formed.

1.6.1 INDUCTION OF ENZYME

Several enzymes are inducible, i.e., they are synthesized only in the presence of inducers. The range of inducers vary from substrate (sucrose, starch, galactosides) to product or any intermediate formed in the course of enzymatic action (fatty acid, phenyl acetate, xylobiose). At times, the inducers are expensive or their handing very precise and difficult; therefore, attempts are being made by researchers to develop mutants of microorganisms in which inducer dependence is completely eliminated.

1.6.2 FEEDBACK REPRESSION

Feedback repression is a phenomenon in which the end product of an enzyme significantly influences the enzyme synthesis, i.e., the entire process, thereby making large-scale production of such regulated enzymes very difficult. Mutants that lack feedback repression have been developed to upregulate the production of enzymes industrially.

1.6.3 NUTRIENT REPRESSION

The native metabolism of microorganisms is devised such that no production of unnecessary enzymes occurs by nutrient repression, which needs to be suppressed for large-scale production. In other words, the microorganisms do not synthesize enzymes that are not required by them, since this is a wasteful exercise in terms of energy expenditure. The inhibition of unwanted enzyme production is done by a technique known as nutrient repression.

A very prominent example of nutrient repression is glucose repression (more appropriately known as catabolite repression). This phenomenon can be explained as follows: in the presence of glucose, the enzymes needed for the metabolism of the rest of the compounds are not synthesized, which is a wasteful exercise in terms of energy. Glucose repression can be overcome by feeding of carbohydrate to the fermentation medium in a way such that the concentration of glucose is almost zero at any given time and is thus technically unavailable. For certain microorganisms, other carbon sources such as pyruvate, lactate, citrate, and succinate also act as catabolite repressors. Nitrogen source repression has also been observed in microorganisms, which may be due to ammonium ions or amino acids. The most commonly inexpensive ammonium salts are used as nitrogen sources.

1.7   GENETIC ENGINEERING OF MICROBIAL CELL FOR ENZYME PRODUCTION

Proteins/enzymes are the functional products of genes. Therefore, theoretically, enzymes are considered to be good candidates for improvement in terms of production through genetic engineering. The advances in the recombinant DNA technology have helped in strain improvement for increasing the microbial production of commercial enzymes. It has even become possible to transfer the desired enzyme genes from one organism to the other organism that has the capability of forming the product (i.e., expression) to a high amount. Production of enzyme has been revolutionized by CRISPR–Cas-mediated genome editing technology that provides a predictable and precise method for genome modification, which can be done in a robust and reproducible fashion. Emergence of system biotechnology and synthetic biology approaches coupled with CRISPR–Cas technology has the ability to change the future of cell factories to possess some new features that are unknown and have not been found naturally [4].

1.8   GENETIC ENGINEERING/PROTEIN ENGINEERING AND CLONING STRATEGIES

The structure of a protein/enzyme can be altered by protein engineering and site-directed mutagenesis to fulfill the following objectives:

- to increase the enzyme stability and catalytic function,
- to generate resistance to oxidation,
- to make altered substrate preference,
- to enhance tolerance to alkali and organic solvents.

Site-directed mutagenesis and protein engineering lead to change in selected amino acids at specific positions (in enzyme) for the production of an enzyme with desired properties. Protein engineering has been used to structurally modify phospholipase A_2 that has the ability to resist high concentration of acid.

Along with this, another approach that is used on a wide scale to enhance the industrial production of an enzyme is cloning the desired gene and transforming a high-yielding variety of microorganisms with the cloned gene.

A diagrammatic representation of a cloning strategy for industrial production of enzymes is given in Figure 1.2. This involves the development of a cDNA library for the mRNA and creation of oligonucleotide probes for the desired enzyme. On hybridization with oligonucleotide probes, the specific cDNA clones can be identified and used. Transformation of industrially important host organism for the production of the desired enzyme is the next objective for which cloning can be employed as given in examples discussed below:

1. The enzyme lipolase, found in the fungus *Humicola languinosa*, is very effective to remove fat stains in fabrics and has been produced using this approach. Indigenous production of this enzyme occurs at a very low concentration; therefore, to overcome this shortcoming, the gene responsible for lipolase was isolated, cloned, and inserted into *Aspergillus oryzae*.

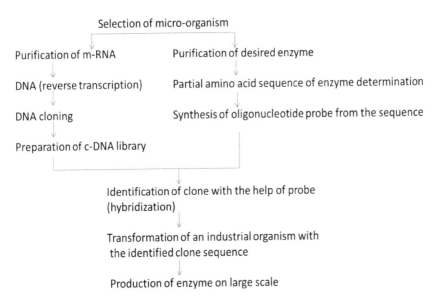

FIGURE 1.2 Production of enzyme aided by genetic engineering and cloning strategy: the gene of interest is reverse transcribed from the isolated m-RNAs by using reverse transcriptase and used in the formation of gene library. The gene of interest is identified with the help of probe synthesized from the desired protein. The identified gene is used to transform a microbial cell, which is used for the industrial preparation of the enzyme.

2. Rennet (chymosin) is an enzyme widely used in making cheese and thus has great commercial importance. Stomachs of young calves used to be its only source. But as it is known, the production, extraction, and purification of an enzyme from an animal source has multiple shortcomings, leading to shortage in its supply. The gene for the synthesis of chymosin has been cloned in microorganisms for its large-scale production, thus bypassing the animal source.

1.9 ENZYMES, CLASSIFICATION, AND THEIR USE

Because enzymes are selective for their substrates and speed up only a few reactions from among the many possibilities, the set of enzymes made in a cell determines which metabolic pathways occur in that cell. Enzymes are very specific because both the enzyme and the substrate possess specific complementary geometric shapes that fit exactly into one another as stated by the Nobel laureate Emil Fischer in 1894 (Figure 1.3), often referred to as "the lock and key" model. However, while this model explains enzyme specificity, it fails to explain the stabilization of the transition state that enzymes achieve.

Enzymes are mostly larger complexes than the substrates they act upon. A very small portion of the enzyme (around 2–4 amino acids) is directly involved in catalysis; it is known as the active site that contains catalytic residues, binds the substrate, and then carries out the reaction.

Enzymes can also contain sites that bind cofactors, which are needed for catalysis. Some enzymes also have binding sites for small molecules, which are often direct or indirect products or substrates of the reaction catalyzed. This binding can serve to increase or decrease the enzyme's activity, providing a means for feedback regulation.

Enzymes are long, linear chains of amino acids that fold to produce a three-dimensional product as the other proteins. Each unique amino acid sequence produces a specific structure, which has unique properties. Individual protein chains may sometimes group together to form a protein complex.

Most enzymes can be denatured (reversibly or irreversibly) and inactivated by processes such as heating or using chemical denaturants, which

Lock and Key mechanism

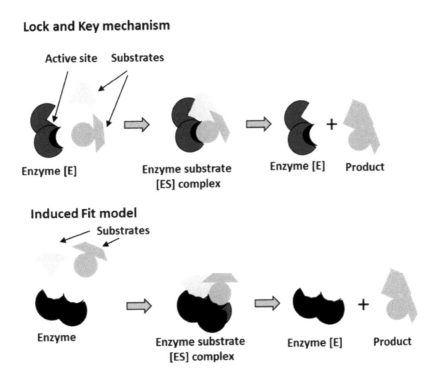

FIGURE 1.3 Model of enzyme action (a) lock and key model having exact fit, (b) induced fit model.

can disrupt the three-dimensional structure of the protein, hence deactivating the enzyme. Enzymes have been classified according to their enzyme catalyzing reaction as shown in Table 1.5.

The enzyme nomenclature scheme was developed starting in 1955, when the International Congress of Biochemistry in Brussels sets up an Enzyme Commission. The first version was published in 1961. The current sixth edition, published by the International Union of Biochemistry and Molecular Biology in 1992, contains 3196 different enzymes. The International Union of Biochemistry (IUB) initiated standards of enzyme nomenclature, which recommend that enzyme names indicate both the substrate acted upon and the type of reaction catalyzed. The Enzyme Commission number (EC number) is a numerical classification scheme for enzymes,

TABLE 1.5   Examples of Few Classes of Industrially Important Enzymes

S. No	Enzyme class (EC number)
1.	Oxidoreductase (EC 1)
2.	Hydrolase (EC 3)
3.	Transferase (EC 2)
4.	Lyase (EC 4)
5.	Isomerase (EC 5)
6.	Ligase (EC 6).

based on the chemical reactions they catalyze, and every EC number is associated with a recommended name for the respective enzyme. Except for some of the originally studied enzymes such as pepsin, rennin, and trypsin, most enzyme names end in "ase."

According to the enzyme commission, the enzymes are broadly classified into six categories.

1.10   ENZYME KINETICS

1.10.1   HOW ENZYMES WORK?

The enzymatic catalysis of reactions is essential to living systems. Under biologically relevant conditions, uncatalyzed reactions tend to be slow. Reactions required to digest food, send nerve signals, or contract a muscle, which are very important for a live cell, do not occur at a useful rate, and without catalysis, most biological molecules are quite stable in the neutral-pH, mild temperature, aqueous environment inside cells. Furthermore, many common reactions in biochemistry entail chemical events that are unfavorable or unlikely in the cellular environment, such as the transient formation of unstable charged intermediates or the collision of two or more molecules in the precise orientation required for reaction.

An enzyme circumvents these problems by providing a specific environment within which a given reaction can occur more rapidly; an enzyme-catalyzed reaction occurs within the confines of a pocket on the enzyme called the active site (Figure 1.3). The molecule that is bound in the active site and acted upon by the enzyme is called the substrate. The surface of the active site is lined with amino acid residues with substituent groups that bind the substrate and catalyze its chemical transformation.

The enzyme substrate complex, whose existence was first proposed by Charles-Adolphe Wurtz in 1880, is central to the action of enzymes. It is also the starting point for mathematical treatments that define the kinetic behavior of enzyme-catalyzed reactions and for theoretical descriptions of enzyme mechanisms.

1.10.2 ENZYMES AFFECT REACTION RATES, NOT EQUILIBRIA

A simple enzymatic reaction might be written as:

$$E + S \rightarrow ES \rightarrow EP \rightarrow E + P$$

where E, S, and P represent the enzyme, substrate, and product; ES and EP are transient complexes of the enzyme with the substrate and with the product.

To understand catalysis, the important distinction between reaction equilibria and reaction rates should be made clear. The function of a catalyst is to increase the *rate* of a reaction. Catalysts do not affect reaction *equilibria*. Any reaction, such as [S] to [P], can be described by a reaction coordinate diagram (Figure 1.4), a picture of the energy changes during the reaction. Energy in biological systems is described in terms of free energy, G.

In the coordinate diagram, the free energy of the system is plotted against the progress of the reaction (the reaction coordinate). The starting point for either the forward or the reverse reaction is called the ground state and the contribution to the free energy of the system by an average molecule (S or P) under a given set of conditions.

a. The lock and key model of enzyme action on substrate postulates an exact fit between the active site of enzyme and substrate.

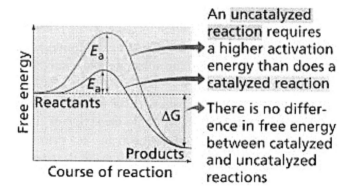

FIGURE 1.4   The free energy of the system is plotted against the progress of the reaction showing the energy changes during the reaction (y-axis), and the horizontal axis shows the progressive chemical changes as the substrate is converted to product.

b. The induced fit model illustrates the fact that a rough initial fit between active site and substrate that alters the active site to its catalytic form.

1.10.3 ENZYME KINETICS: MODEL OF ENZYME ACTION

$$E+S \underset{k_{-1}}{\overset{k_1}{\rightleftharpoons}} ES \xrightarrow{k_{cat}} E+P \tag{1}$$

In this model, the substrate S reversibly associates with the enzyme E in a first step, and some of the resulting complex ES breaks down and yields the product P and the free enzyme back.

For this model, let v_0 be the initial velocity of the reaction. The latter stands for the appearance of the product P in solution ($+ d[P]/dt$) whose phenomenological rate equation (first-order) is given by

$$v_0 = k_{cat}[ES] \tag{2},$$

containing an experimentally measurable (dependent) variable – v_0, a kinetic parameter – k_{cat}, and another variable unknown to us – [ES].

Before proceeding, one should state (and remember) some implicit assumptions:

- As long as initial velocity is considered, the concentration of product can be neglected (compared to that of the substrate, thus [P] << [S]), and
- The concentration of substrate is in large excess over that of the enzyme ([E] << [S]).

These assumptions, which hold in most kinetic experiments performed in test tubes at low enzyme concentration, are convenient when considering the mass conservation equations for the reactants $[S]_0 = [S]_{free} + [ES] + [P]$ which now approximates to $[S]_0 = [S]$, while that for the enzyme is $[E]_{total} = [E]_{free} + [ES]$ (the possible formation of a complex EP is not considered here).

We want to express v_0 in terms of measurable (experimentally defined, independent) variables, like [S] and $[E]_{total}$. So, we must replace the unknown [ES] in (2) with measurables.

During the initial phase of the reaction, as long as the reaction velocity remains constant, the reaction is in a steady state, with ES being formed and consumed at the same rate. During this phase, the rate of formation of [ES] (one second order kinetic step) equals its rate of consumption (two first order kinetic steps). According to model (1), Rate of formation of $[ES] = k_1[E][S]$, and rate of consumption of $[ES] = k_{-1}[ES] + k_{cat}[ES]$.

So, in the steady state,

$$k_{-1}[ES] + k_{cat}[ES] = k_1[E][S] \tag{3}$$

to solve for [ES] can replace in (2). First, collect the kinetic constants, and the concentrations (variables) in (3):

$$(k_{-1} + k_{cat})\,[ES] = k_1\,[E][S],$$

and

$$(k_{-1} + k_{cat})/k_1 = [E][S]/[ES] \tag{4}$$

To simplify (4), first group the kinetic constants by defining them as K_m:

$$K_m = (k_{-1} + k_{cat})/k_1 \tag{5}$$

and then express [E] in terms of [ES] and $[E]_{total}$, to limit the number of unknowns:

$$[E] = [E]_{total} - [ES] \tag{6}$$

Substitute (5) and (6) into (4):

$$K_m = ([E]_{total} - [ES]) [S]/[ES] \tag{7}$$

To solve (7) for [ES], first multiply both sides by [ES]:

$$[ES] K_m = [E]_{total}[S] - [ES][S]$$

Then, collect terms containing [ES] on the left:

$$[ES] K_m + [ES][S] = [E]_{total}[S]$$

Factor [ES] from the left-hand terms:

$$[ES](K_m + [S]) = [E]_{total}[S]$$

and finally, divide both sides by $(K_m + [S])$:

$$[ES] = [E]_{total} [S]/(K_m + [S]) \tag{8}$$

Substitute (8) into (2):

$$v_0 = k_{cat}[E]_{total} [S]/(K_m + [S]) \tag{9}$$

The maximum velocity V_{max} occurs when the enzyme is saturated — that is, when all enzyme molecules are tied up with S, or $[ES] = [E]_{total}$. Thus,

$$V_{max} = k_{cat} [E]_{total} \tag{10}$$

Substitute V_{max} into (9) for $k_{cat} [E]_{total}$:

$$v_0 = V_{max} [S]/(K_m + [S]) \qquad (11)$$

This equation expresses the initial rate of reaction in terms of a measurable quantity, the initial substrate concentration. The two kinetic parameters, V_{max} and K_m, will be different for every enzyme-substrate pair.

Equation (11), the Michaelis-Menten equation, describes the kinetic behavior of an enzyme that acts according to the simple model (1).

Mathematically, the function v_0 presents two asymptotes:

- one parallel to the [S] axis at $v_0 = V_{max}$, represents the velocity at infinite [S] (saturation),
- the second parallel to the v_0 axis at $[S] = -K_m$, has no physical meaning (no negative concentrations).

Further analysis reveals the physical meaning of K_m: the concentration of substrate at which the velocity is half V_{max}. Indeed, substituting K_m for [S] in (11) yields

$$v_0 = 1/2 \, V_{max}$$

Thus, a low value for K_m may indicate a high affinity of the enzyme for its substrate.

Another physically meaningful limit of this function is found at vanishingly small values of [S] (--> 0), where

$$v_0 \rightarrow V_{max}/K_m \, [S]$$

In this case, the velocity becomes proportional to the (low, relative to K_m) substrate concentration, displaying pseudo-first order kinetics in [S].

The parameter V_{max}/K_m (or rather its constant part k_{cat}/K_m), often referred to as the catalytic ability of the enzyme, is a direct measure of the efficiency of the enzyme in transforming the substrate S.

k_{cat}/K_m recombines the two traditionally separated aspects of enzyme catalysis:

- the effectiveness of transformation of bound product (catalysis *per se*, k_{cat});
- the effectiveness of productive substrate binding (affinity, $1/K_m = k_1/$

$(k_{-1} + k_{cat})).$

Equation (11) means that, for an enzyme acting according to the simple model (1), a plot of v_0 versus [S] will be a rectangular hyperbola. When enzymes exhibit this kinetic behavior, unless we find other evidence to the contrary, we assume that they act according to model (1) and call them Michaelis–Menten enzymes.

Michaelis and Menten's lasting contribution to enzymology has played a fundamental role in understanding enzyme biochemistry in the test tube.

The Michaelis–Menten equation assumes that the number of enzyme and substrate molecules is macroscopically large [5, 6]. This is a fundamentally limiting assumption when one considers that the number of molecules of many chemical species inside cells ranges from tens to a few thousands [7, 8], a number many orders of magnitude smaller than that in typical test tube experiments.

This intrinsic noise stems from the random timing of biochemical reaction events. The randomness has various sources of origin including the Brownian motion of reactants [9]. During the last 20 years, the development of mathematical and computational approaches to investigate the inherent stochasticity of reactions inside the cell has been propelled by advances in experimental techniques that are capable of following reactions at the single-molecule level by using fluorescence microscopy and related optical methods [8, 10–17], thus changing and adding many other dimensions to Michaelis and Menten's contribution [18].

1.11   NUCLEIC ACID CATALYST

Enzymes are mostly proteinaceous in nature, but the discovery of self-cleavage and ligation activity of the group I intron after the transcription and before translation has led to the expansion of research interest in catalytic nucleic acids. This knowledge that nucleic acids can act as a catalyst in several reactions has enabled the researchers with a valuable nonprotein resource for the manipulation of biomolecules.

The plethora of reaction in which such catalyst can be used is for gene correction. This application is based on the trans-splicing activity of group I intron. Besides this, we are also aware of the catalytic activity of RNA.

1.11.1   RIBOZYMES

RNA showing catalytic activity has been termed "ribozyme." Ribozymes have great application in mediating gene inactivation as they induce specific RNA cleavage from a very small catalytic domain.

1.11.2   DNA ENZYMES OR DEOXYRIBOZYME

Recently, a new class of catalytic nucleic acid made entirely of DNA has emerged through in vitro selection. DNA enzymes or deoxyribozyme with RNA cleavage activity has demonstrated its capacity for gene suppression both in vitro and in vivo. These new molecules, although rivaling the activity and stability of synthetic ribozymes, are limited by inefficient delivery to the intracellular target RNA. The challenge of in vivo delivery is being addressed with the assessment of a variety of approaches in animal models with the aim of bringing these compounds closer to the clinic.

Nucleic acids often serve as the information storage and transfer molecules for living organisms, and they also participate in a broad variety of other cellular functions. These include amino acid biosynthesis, protein localization, transcriptional control, and translation. Possibly, the most dramatic example of an alternative function for a nucleic acid is the role that RNA plays as a biocatalyst, including RNA molecules that participate in RNA processing and protein synthesis. The chemical interactions that facilitate these biological functions provide a wealth of new research opportunities for researchers working at the diminishing interface of chemistry and biology.

1.12   REVIEW QUESTIONS

1. What are the various sources of enzymes?
2. How can enzymes be produced commercially using a microorganism?
3. List the various tools to enhance the production of enzymes.
4. Why are enzymes from microbes preferred?

KEYWORDS

- activation energy
- enzymes
- enzyme kinetics

REFERENCES

1. Boyer, P. D., (1971). *The Enzymes,* 3rd ed., Academic Press Inc., New York, 5, 115–182.
2. Fersht, A., (2007). Enzyme structure and mechanism, the comprehensive enzyme information system, San Francisco, Brenda, W. H., 50–52, ISBN 0-7167-1615-1.
3. Duffy, J. I., (1980). *Chemicals by Enzymatic and Microbial Processes,* Noyes Data Corp., New Jersey, 368–373.
4. Roointan, A., & Morovat, M. H., (2017). Road to the future of systems biotechnology: CRISPR Cas-mediated metabolic engineering for recombinant protein production. *Biotechnol. Genet. Eng. Rev., 4,* 1–18.
5. Turner, T. E., Schnell, S., & Burrage, K., (2004). Stochastic approaches for modeling *in vivo* reactions. *Comput. Biol. Chem., 28,* 165–178.
6. Grima, R., & Schnell, S., (2008). Modelling reaction kinetics inside cells. *Essays. Biochem., 45,* 41.
7. Ishihama, Y., Schmidt, T., Rappsilber, J., et al., (2008). Protein abundance profiling of the Escherichia coli cytosol. *BMC Genomics., 9,* 102.
8. Walter, N. G., Huang, C. Y., Manzo, A. J., & Sobhy, M. A., (2008). Do-it-yourself guide: how to use the molecule toolkit. *Nat. Methods, 5,* 475–489.
9. Bartholomay, A. F., (1962). A stochastic approach to statistical kinetics with application to enzyme kinetics. *Biochemistry, 1,* 223–230.
10. Xie, X. S., & Trautman, J. K., (1998). Optical studies of single molecules at room temperature. *Annu. Rev. Phys. Chem., 49,* 441–480.
11. Grohmann, D., Werner F., & Tinnefeld, P., (2003). Making connections-strategies for single molecule fluorescence biophysics. *Curr. Opin. Chem. Biol., 17,* 691–698.
12. Coelho, M., Maghelli, N., & Toli, N. I. M., (2013). Single-molecule imaging in vivo: the dancing building blocks of the cell. *Integr. Biol., 5,* 748–758.
13. Xia, T., Li, N., & Fang, X., (2013). Single-molecule fluorescence imaging in living cells. *Annu. Rev. Phys. Chem., 64,* 459–480.
14. Puchner, E. M., & Gaub, H. E., (2012). Single-molecule mechanoenzymatics, *Annu. Rev. Biophys., 41,* 497–518.
15. Elson, E. L., (2011). Fluorescence correlation spectroscopy: past, present, future. *Biophys. J., 101,* 2855–2870.

16. Harriman, O. L. J., & Leake, M. C., (2011). Single molecule experimentation in biological physics: exploring the living component of soft condensed matter one molecule at a time. *J. Phys.: Condens. Matter, 23*, 503101.

17. Li, G. W., & Xie, X. S., (2011). Central dogma at the single molecule level in living cells. *Nature, 475*, 308–315.

18. Grima, R., Walter N. G., & Santiago, S., (2014). Single-molecule enzymology a la Michaelis–Menten. *FEBS J., 281*, 518–530.

CHAPTER 2

IMMOBILIZATION BIOCATALYSIS

ANJALI PRIYADARSHINI and PRERNA PANDEY

CONTENTS

2.1 INTRODUCTION

Enzymes are catalyst that catalyze various biochemical and chemical reactions. They are universally present in plants and animals. The enzymes are catalysts work at a specific optimum temperature, pressure, pH, and substrate specificity for the production of desired products without any intermediate products as contaminants. Thus, there is an increasing demand for variety of enzymes for use in:

i. Food industries – such as baking, dairy product, starch conversion, and beverage processing (fruit, vegetable juices, beer, and wine), etc.
ii. Textile industries.
iii. In paper and pulp making and detergents, the use of enzymes has become a necessary processing strategy.
iv. Healthcare and pharmaceuticals and chemical manufacturing have increased due to the catalytic nature of enzymes.
v. Waste management especially for solid waste treatment and waste water purification.
vi. Production of biofuels such as biodiesel, bioethanol, biohydrogen, and biogas from biomass conversion.

However, all these desirable characteristics of enzymes and their widespread industrial applications are often hindered due to lack of long-term operational stability, low shelf life, and their recovery and reusability after product formation. Thus, enzyme immobilization has emerged as one of the strategies to overcome or reduce these problems.

2.2   ENZYME IMMOBILIZATION

Enzyme immobilization can be defined as a process in which the movement of a molecule is restricted in space completely or to a small limited region aided by its attachment to a solid structure. Immobilization has been observed in many natural processes such as those existing in the form of biofilms formed by microorganisms on variety of surfaces, for example, a rock immersed in a stream, an implant in the human body, a tooth, a water pipe or conduit, etc. are all sites where biofilms develop [1]. Many molecules have been immobilized, and the majority of them are biomolecules due to their biological and biomedical applications. The following are examples of some of these molecules:

• Proteins: Enzymes, antibodies, antigens, cell adhesion molecules, and "blocking" proteins.
• Drugs: Anticancer agents, antithrombogenic agents, antibiotics, contraceptives, drug antagonists, and peptide/protein drugs.
• Saccharides: Sugars, oligosaccharides, and polysaccharides.

- Lipids: Fatty acids, phospholipids, glycolipids, and any fat-like substances.
- Ligands: Hormone receptors, cell surface receptors, avidin and biotin – in immunology, small molecules that are bound to another chemical group or molecule.
- Nucleic acids and nucleotides: DNA, RNA – High-molecular-weight (MW) substances formed of sugars, phosphoric acid, and nitrogen bases (purines and pyrimidines).
- Others: Conjugates or mixtures of any of the above.

2.2.1   ADVANTAGES OF IMMOBILIZED ENZYME

- The ability to stop the reaction rapidly by removing the enzyme from the reaction solution or initiate the reaction by adding the enzyme.
- Easy separation of enzyme from the product (especially useful in food and pharmaceutical industries where minutest contamination of the product can be detrimental).
- Immobilization provides a physical support for enzymes, cells, and other molecules, thus increase in the stability of enzyme against pH, temperature, solvents, contaminants, and impurities. Immobilization of enzymes is one of the main methods used to stabilize free enzymes [2].

2.3   METHODS OF IMMOBILIZATION

Immobilized enzyme was discovered in 1916 [3]. It was demonstrated that the activity of invertase enzyme is not affected when it is adsorbed on a solid matrix such as charcoal or aluminum hydroxide. This aspect led to the development of currently available various enzyme immobilization techniques, which have been broadly categorized as follows (Figure 2.1):

1. Adsorption onto an inert carrier.
2. Entrapment within the lattice of a polymerized gel (synthetic and nonsynthetic).
3. Cross-linking of the protein with a bifunctional reagent.
4. Covalent bonding to a reactive insoluble support.

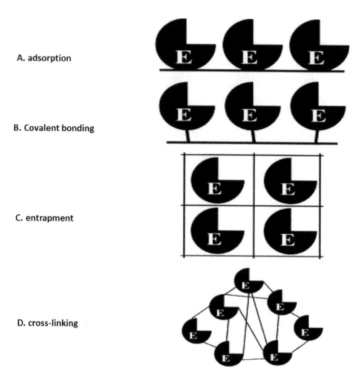

FIGURE 2.1 Diagrammatic representation of the various methods of enzyme immobilization.

2.3.1 CHOICE OF IMMOBILIZATION METHOD

When immobilizing an enzyme on a surface, it is most important to choose a method of attachment aimed to reactive groups outside the active catalytic and binding site of that enzyme. Considerable knowledge of active sites of particular enzymes will enable methods to be chosen that would avoid reaction with the essential group therein. Alternatively, these active sites can be protected during attachment as long as the protective groups can be removed without any loss of enzyme activity. In some cases, this protective function can be fulfilled by a substrate of the enzyme or a competitive inhibitor; this also contributes toward retention of tertiary structure of the enzyme.

The surface on which the enzyme is immobilized has several vital roles to play, such as retaining of tertiary structure in the enzyme by hydrogen bonding or the formation of electron transition of tertiary structure in the enzyme by hydrogen bonding or the formation of electron transition com-plexes. Retention of tertiary structure may also be a vital factor in maxi-mizing thermal stability in the immobilized state. In this respect, it is wise to follow closely the new findings in the chemical nature of soluble ther-mostable enzyme. The microenvironment of surface and the immobilized enzyme has an anionic or cationic nature of the surface that can cause a dis-placement at the optimum pH of the enzyme up to 2 pH units. This may be accompanied by a general broadening of the pH region in which the enzyme can work effectively. Immobilization by cross-linking the protein enzyme in order to insolubilize it or merely immobilize it in the desired location has many possibilities and is relatively cheap. Several aldehydes and other cross-linking agents are now available for this purpose. Extension of this approach to a process where an enzyme is an integral component of a copo-lymer could permit designing of reversible polymerization. Therefore, the immobilization method must be designed to minimize enzyme desorption, to maximize the stability of enzyme on the support, and to maximize the access of the substrate to the active site of the enzyme.

2.3.2 ADSORPTION

Immobilization by adsorption is among the simplest method and includes surface interactions between enzyme and support material, which are reversible as shown in Figure 2.2. The procedure of adsorption consists of mixing together the biological component(s) and a support with adsorp-tion properties, under suitable conditions of pH and ionic strength for a period of incubation, followed by collection of the immobilized material and extensive washing to remove the unbound biological components. Adsorption involves the physical binding of enzymes (or cells) on the sur-face of an inert support. The support materials may be:

a)   inorganic (e.g., alumina, silica gel, calcium phosphate gel, glass)
b)   organic (starch, carboxymethyl cellulose, DEAE-cellulose, DEAE-sephadex).

The first industrially used immobilized enzyme was prepared by adsorption of amino acid acylase on DEAE-cellulose [4, 5] which reported the role of hydrophobic surfaces of nanoporous silica glasses on protein folding enhancement.

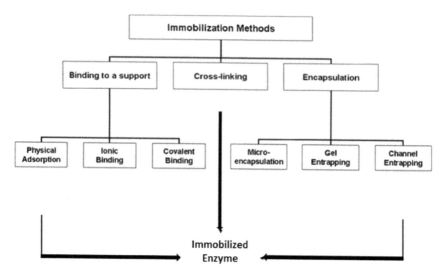

FIGURE 2.2   Different methods for enzyme immobilization.

Advantages of enzymes immobilized using the adsorption technique:

a) Reversibility, which enables not only the purification of proteins but also the reuse of the carriers.
b) Simplicity, which enables enzyme immobilization under mild conditions.
c) Possible high retention of activity because there is no chemical modification [6].
d) Cheap and quick method.
e) No chemical changes to the support or enzyme occur.

Disadvantages of enzymes immobilized using the adsorption technique:

a) Adsorption of enzyme molecules (on the inert support) involves weak forces such as van der Waals forces and hydrogen bonds. Therefore, the adsorbed enzymes can be easily removed by minor changes in pH, ionic strength, or temperature. This is a disadvantage for industrial use of enzymes.
b) Contamination of product.

c) Nonspecific binding.
d) Overloading on the support.
e) Steric hindrance by the support.

Consequently, a number of variations have been developed in recent decades to solve this intrinsic drawback. Examples are adsorption–cross-linking; modification–adsorption; selective adsorption–covalent attachment; adsorption–coating, etc.

2.3.3 COVALENT BINDING

Immobilization of the enzymes can be achieved by creation of covalent bonds between the chemical groups of enzymes and the chemical groups of the support (Figure 2.3). This technique is widely used. However, covalent binding is often associated with loss of some enzyme activity. The inert support usually requires pretreatment (to form pre-activated support) before it binds to enzyme. The following are the common methods of covalent binding. Immobilization of enzymes by their covalent coupling to insoluble matrices is an extensively researched technique. Only small amounts of enzymes may be immobilized by this method (about 0.02 g per gram of matrix) although in exceptional cases as much as 0.3 g per gram of matrix has been reported. The strength of binding is very strong, however, and very little leakage of enzyme from the support occurs. The relative usefulness of various groups, found in enzymes, for covalent link formation depends upon their availability and reactivity (nucleophilicity), in addition to the stability of the covalent link, once formed (Figure 2.3). The reactivity of the protein side-chain nucleophiles is determined by their state of protonation (i.e., charged status) and roughly follows the relationship $-S^- > -SH > -O^- > -NH_2 > -COO^- > -OH >> -NH_3^+$, where the charges may be estimated from the knowledge of the pK_a values of the ionizing groups and the pH of the solution. Lysine residues are found to be the most generally useful groups for covalent bonding of enzymes to insoluble supports due to their widespread surface exposure and high reactivity, especially in slightly alkaline solutions. They also appear to be only very rarely involved in the active sites of enzymes. This method of immobilization involves formation of a covalent bond between the enzyme and

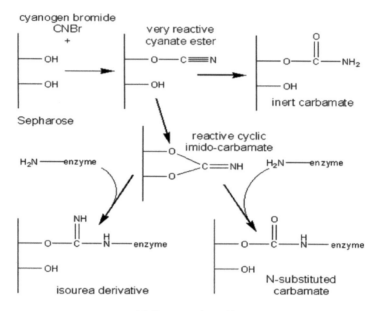

(a) Cyanogen bromide

(b) Ethyl chloroformate

(c) Carbodiimide

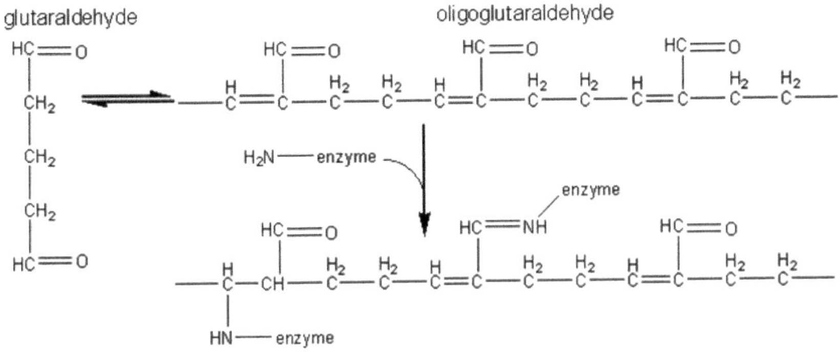

(d) Glutaraldehyde

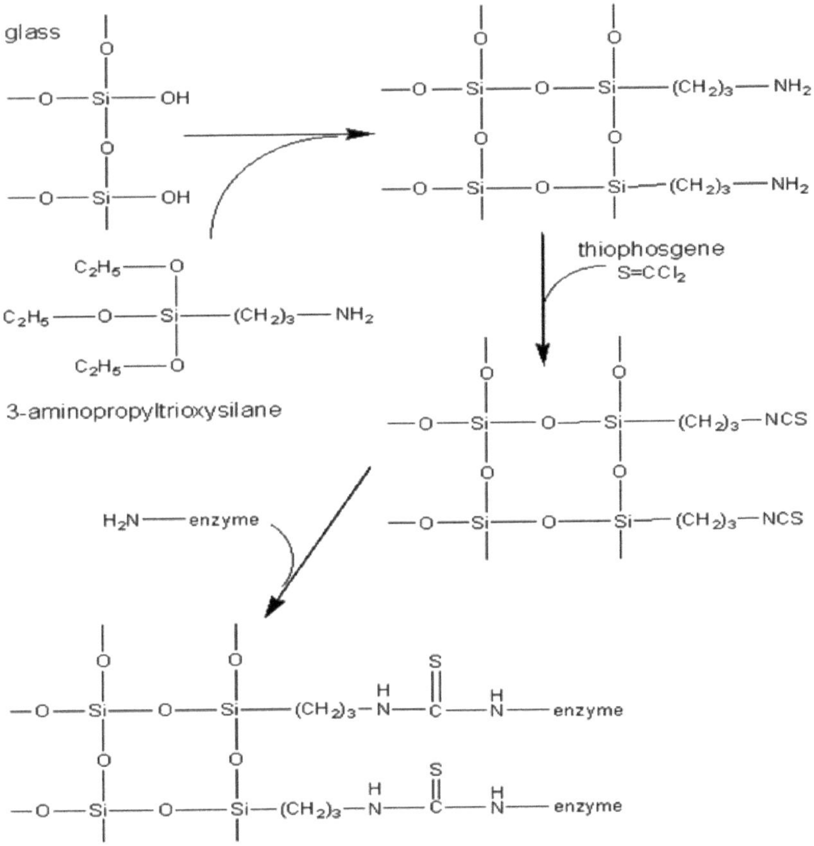

(e) 3-aminopropyltriethoxysilane

FIGURE 2.3 Commonly used methods for the covalent immobilization of enzymes.

support material. Covalent bonds provide the strongest bond between carrier and enzyme, compared with other types of enzyme immobilization methods. Thus, the major advantage with this process is that the problem of leakage of enzyme from the matrix used is often minimized [7]. The bond is normally formed between functional groups present on the surface of the support and functional groups belonging to amino acid residues on the surface of the enzyme. There are many reaction procedures for coupling an enzyme to a support via a covalent; most reactions fall into the following categories:

a) formation of an iso-urea linkage;
b) formation of a diazo linkage; and
c) formation of a peptide bond or an alkylation reaction.

2.3.4   STEPS FOR COVALENT BINDING OF ENZYME TO THE MATRIX

2.3.4.1   Cyanogen Bromide Activation

The inert support materials (cellulose, sepharose, sephadex) containing glycol groups are activated by CNBr, which then bind to enzymes and immobilize them (Figure 2.3A).

2.3.4.2   Diazotation

Some of the support materials (amino benzyl cellulose, amino derivatives of polystyrene, aminosilanized porous glass) are subjected to diazotation on treatment with NaNO$_2$ and HCl. They, in turn, bind covalently to tyrosyl or histidyl groups of enzymes (Figure 2.3B).

2.3.4.3   Peptide Bond Formation

Enzyme immobilization can also be achieved by the formation of peptide bonds between the amino (or carboxyl) groups of the support and the carboxyl (or amino) groups of enzymes (Figure 2.3C). The support material is first chemically treated to form active functional groups.

2.3.4.4   Activation by Bi- or Poly-Functional Reagents

Some of the reagents such as glutaraldehyde can be used to create bonds between the amino groups of enzymes and the amino groups of supports (e.g., aminoethylcellulose, albumin, amino-alkylated porous glass). This is depicted in Figure 2.3D.

Advantages of enzymes immobilized using the covalent technique:

a)   There is no problem of leakage of enzyme.
b)   There is an increase in the thermal stability of the enzyme.
c)   The contact between enzyme and substrate is facilitated due to the localization of enzyme on support materials.

2.3.4.5   Disadvantages of Enzymes Immobilized Using the Covalent Technique

a)   The cost is quite high as good supports are very expensive (e.g., Eupergit C and Agaroses).
b)   Loss of enzyme activity may occur, for example, mismatched orientation of enzyme on the carriers, which results due to involvement of active center in the binding and not free to enhance substrate to product formation.

1.   Activation of Sepharose by cyanogen bromide. Conditions are chosen to minimize the formation of the inert carbamate.
2.   Chloroformates produce intermediates like those of cyanogen bromide sans its toxicity and may be used.
3.   Carbodiimides may be used to attach amino groups on the enzyme to carboxylate groups on the support or carboxylate groups on the enzyme to amino groups on the support. Conditions are chosen to minimize the formation of the inert substituted urea.

4. Glutaraldehyde is used to cross-link enzymes or link them to supports. It usually consists of an equilibrium mixture of monomer and oligomers. The product of the condensation of enzyme and glutaraldehyde may be stabilized against dissociation by reduction with sodium borohydride.
5. The use of trialkoxysilane to derivatize glass. The reactive glass may be linked to enzymes by a number of methods, including the use thiophosgene.

The most commonly used method for immobilizing enzymes includes Sepharose that is activated by cyanogen bromide. Sepharose is a commercially available beaded polymer which is highly hydrophilic and generally inert to microbiological attack. The hydroxyl groups of this polysaccharide combine with cyanogen bromide to give the reactive cyclic imido-carbonate. This reacts with primary amino groups (i.e., mainly lysine residues) on the enzyme under mildly basic conditions (pH 9–11.5, Figure 2.3A). The high toxicity of cyanogen bromide has led to the commercial, if rather expensive, production of ready-activated Sepharose and the investigation of alternative methods, often involving chloroformates, to produce similar intermediates (Figure 2.3B). Carbodiimides (Figure 2.3C) are very useful bifunctional reagents as they allow the coupling of amines to carboxylic acids. Careful control of the reaction conditions and choice of carbodiimide allow a great degree of selectivity in this reaction. Glutaraldehyde is another bifunctional reagent that may be used to cross-link enzymes or link them to supports (Figure 2.3D). It is particularly useful for producing immobilized enzyme membranes, for use in biosensors, by cross-linking the enzyme plus a noncatalytic diluent protein within a porous sheet (e.g., lens tissue paper or nylon net fabric). The use of trialkoxysilanes allows even such apparently inert materials as glass to be coupled to enzymes (Figure 2.3E). There are numerous other methods available for the covalent attachment of enzymes (e.g., the attachment of tyrosine groups through diazo-linkages, and lysine groups through amide formation with acyl chlorides or anhydrides).

2.3.4 ENTRAPMENT

Enzymes can be immobilized by physical entrapment inside a polymer or a gel matrix. The size of the matrix pores is such that the enzyme is retained, while the substrate and product molecules pass through. In this technique, commonly referred to as lattice entrapment, the enzyme (or cell) is not subjected to strong binding forces and structural distortions.

Entrapment can be achieved by mixing an enzyme with a polyionic polymer material, such as carrageenan, and by crosslinking the polymer with multivalent cations, e.g., hexamethylene diamine, in an ion-exchange reaction to form a lattice structure that traps the enzymes; this is termed ionotropic gelation. The various matrices used for entrapment of enzyme includes polyacrylamide gel, gelatin, silicone, rubber, starch, and collagen to name a few. Advantages of enzymes immobilized using the entrapment technique is that the enzyme loading is very high. Different modes of entrapment are inclusion in gel, inclusion in fiber, and inclusion in microcapsule (Figure 2.4A–C)

The major drawback of the entrapment method for enzyme immobilization is their leakage from the matrix, which results in great loss as well as gains to the overall cost of the product. Despite the abovementioned limitation of entrapment procedure, it remains the method of choice for immobilizing whole cells, which are the source of various enzymes and used in a number of fermentation procedures. Entrapped cells are used for industrial production of various amino acids (L-isoleucine, L-aspartic acid), L-malic acid, and hydroquinone to name a few.

2.3.4.1 Microencapsulation

Entrapment of an enzyme by microencapsulation refers to the process of spherical particle formation. In this method, a liquid or suspension is enclosed in a semipermeable membrane (Figure 2.4C). The membrane may be polymeric, lipoidal, lipoprotein-based, or non-ionic in nature. Microencapsulation is achieved in special membrane reactors where emulsions are formed and stabilized to form microcapsules. The semipermeable membrane hinders the passage of enzyme, which is usually a large molecule but allows the passage of substrate and product through it. Thus, this type of entrapment is beneficial in the sense that the microcapsules can be harvested after the completion of the process and reused, as enzymes are biocatalyst that do not undergo chemical change during the process.

Disadvantages of enzymes immobilized using the entrapment technique:

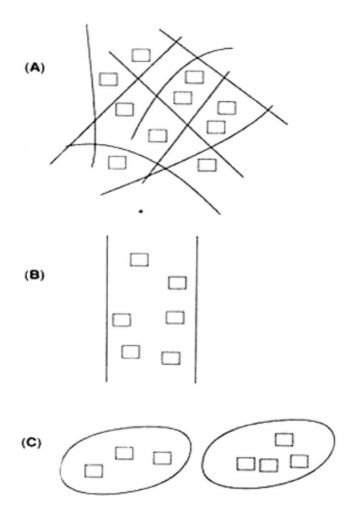

FIGURE 2.4 Immobilization of enzyme using different methods: (A) entrapment by inclusion of enzyme in gel matrix, (B) entrapment of enzyme in fibers, and (C) entrapment by inclusion in microcapsules.

a) Enzyme leakage from the support.
b) Diffusion of the substrate to the enzyme and of the product away from the enzyme (diffusion limitation).
c) The problems associated with diffusion are acute and may result in rupture of the membrane if products from a reaction accumulate rapidly.

2.3.5   ENCAPSULATION

Encapsulation of enzymes as shown in Figure 2.5 can be achieved by enveloping the biological components within various forms of semiper-meable membranes [8]. It is similar to entrapment in that the enzyme is free in solution, but restricted in space. Large proteins or enzymes cannot pass out of, or into the capsule, but small substrates and products can pass freely across the semipermeable membrane. Many materials have been used to construct microcapsules varying from 10–100 μm in diameter. For example, nylon and cellulose nitrate have proven to be popular. Ionotropic gelation of alginates has proven it efficacy for encapsulation of drugs, enzymes, and cells [9]. On the nano scale level, silica-based nanoporous sol-gel glasses for encapsulation and stabilization of some proteins has been used. Advantages of enzymes immobilized using the encapsulation technique:

a) The enzymes could be encapsulated inside the cell.
b) Possibility of coimmobilization, where cells and/or enzymes may be immobilized in any desired combination to suit particular appli-cations.

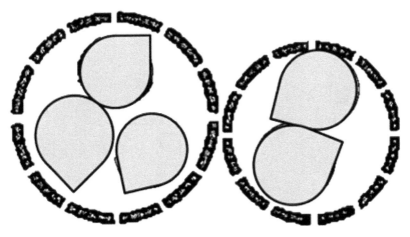

FIGURE 2.5 Immobilization of enzyme using the encapsulation technique in a semipermeable membrane.

2.3.6   CROSS-LINKING

The characteristic of this type of immobilization is support-free as shown in Figure 2.6 and involves joining enzyme molecules to each other to form a large, three-dimensional complex structure, and can be achieved by chemical or physical methods [10].

Chemical methods involve covalent bond formation between the enzymes occurring by means of a bi- or multifunctional reagent, for example, glutaraldehyde, dicarboxylic acid or toluene di-isocyanate. Physical method involves flocculating agents such as polyamines, polyethyleneimine, polystyrene sulfonates, and various phosphates to form crosslink cells using physical bonds.

The limitation of crosslinking is that it provides poor mechanical properties to the aggregates. This is the reason why crosslinking is most often used to enhance the other methods of immobilization already described.

Advantages of enzymes immobilized using the crosslinking technique:

i.   The immobilization is support-free.
ii. Cross-linking between the same enzyme molecules stabilizes the enzymes by increasing the rigidity of the structure

Disadvantages of enzymes immobilized using the crosslinking technique:

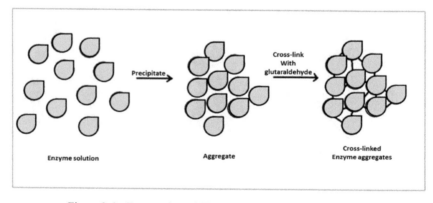

Figure 2.6   Enzyme immobilization by cross-linking technique.

i. Harshness of reagents of crosslinking is a limiting factor in applying this method to many enzymes.
ii. The enzyme may partially lose activity or become totally inactivated in case the cross-linking reagent reacted across the active site.

Matrices for immobilization can be classified according to their chemical composition as organic and inorganic supports (Table 2.1).

The former can be further classified into natural and synthetic matrices as in Table 2.1 [27]. The shape of the carrier can be classified into two types, i.e., irregular and regular shapes such as (A): beads; (B): fibers; (C): hollow spheres; (D): thin films; (E): discs, and (F): membranes. Selection of the geometric properties for an immobilized molecule is largely dependent on the peculiarity of certain applications.

2.4 NEW CARRIERS USED IN IMMOBILIZATION

Over the last few years, mesoporous support such as silica and silicates with pore size of 2–50 nm has been developed and considered as one of the most promising carriers for enzyme immobilization [11–15]. The exploitation of novel carriers that enable high enzyme loading and activity

TABLE 2.1 List of Matrixes Involved in Various Immobilization Techniques

Method	Support material	Cells	Reaction
Adsorption	Gelatin	*Lactobacilli*	Lactose→lactic acid
	DEAE Cellulose	*Nocardia*	Steroid conversion
	Cotton fibers	*Zymomonas*	Glucose →ethanol
	Porous glass	*Saccharomyces*	Glucose→ ethanol
Covalent bonding	Cellulose	*S. cerevisiae*	Glucose →ethanol
	Titanium oxide	*Acetobacter*	Vinegar
Cross linking	Glutaraldehyde	*E. coli*	Fumaric acid
Entrapment	Aluminium alginate	*Candida tropicalis*	Phenol degradation
	Calcium alginate	*S. cerevisiae*	Glucose →ethanol
Encapsulation	Polyester	*Streptomyces* sps	Glucose →fructose

retention has become the focus of recent attention [16]. The large surface areas and greater pore volumes of these materials could enhance the loading capacity of an enzyme and the large pores in the support facilitate transport of substrate and product [17]. Functional mesoporous material resulted in exceptionally high immobilization efficiency with enhanced stability, while conventional approaches yielded far lower immobilization efficiency [18]. Additionally, the increase in the thermal stability of immobilized enzyme indicated that protein inside a confined space could be stabilized by some folding forces which did not exist in proteins in bulk solutions [19]. Confinement of the support nanopore could be similar to the macromolecular crowding [20], and could also stabilize the enzyme at high temperature. Nanoporous gold [21] and nanotube [22, 23] have also been used to immobilize enzymes. Most of the obtained immobilized enzymes were used in the electrode preparation and biosensor applications. The modified porous gold electrode shows an overall increased signal, and therefore, a better detection limit and higher sensitivity when used as sensors.

Magnetic Hybrid Support- The use of magnetic supports for enzyme immobilization enables rapid separation in an easily stabilized fluidized bed reactor for continuous operation of an enzyme. It can also reduce the capital and operation costs [24]. Due to the functionalization [25] of enzyme and its suitable microenvironment, magnetic materials were often embedded in organic polymer or inorganic silica to form hybrid support [26]. Recently, because of the low enzyme loading on the conventional magnetic beads, further attention was paid to the magnetic mesoporous support [27]. Magnetite mesoporous silica hybrid support was fabricated by the incorporation of magnetite to the hollow mesoporous silica shells, which resulted in the perfect combination of mesoporous materials properties with magnetic property. The produced hybrid support has shown to improve enzyme immobilization [27].

2.5 NANOPARTICLES AS IMMOBILIZATION MATRIX

Nanoparticles act as very efficient support materials for enzyme immobilization, because of their ideal characteristics for balancing the key factors that determine biocatalyst efficiency, including specific surface area, mass transfer resistance, and effective enzyme loading [29–32]. Diffusion problem is more relevant when we are dealing with the macromolecular substrates; the nanoparticles are the ideal candidates for such systems [33]. Moreover, the enzyme-bound nanoparticles show Brownian movement when dispersed in aqueous solutions, showing that the enzymatic activities are comparatively better than that of the unbound enzyme [34]. In addition, magnetic nanoparticles possess additional advantage, as they can be separated easily using an external magnetic field. Enzymatic immobilization on Au and Ag nanoparticles have been studied using either as whole cells or isolated enzymes, which include lysozyme, glucose oxidase, aminopeptidase, and alcohol dehydrogenase [35–38].

2.6 APPLICATIONS OF IMMOBILIZED MOLECULES

2.6.1 DRUG DELIVERY SYSTEMS

Advanced drug delivery systems (ADDS) have found applications in many biomedical fields. Drug delivery field combines material science, pharmaceutics, and biology [39]. Adoption of different types of membranes in ADDS has enabled to release drug in an optimal fashion according to the nature of a disease [40]. Examples of drug delivery systems are: glucose-sensitive insulin and drug-loaded magnetic nanoparticles. Magnetic nanoparticles loaded with drugs provides the means to send the drugs to targeted sites and also has the added advantage of drug released in a controlled manner. This property of controlled drug release reduces the side effects due to lowering of dosage and minimizing or preventing degradation of drug using pathways other than gastrointestinal. Drug-loaded magnetic nanoparticles (DLMNP) have several advantages such as: small particle size; large surface area; magnetic response; biocompatibility, and nontoxicity. DLMNP is introduced through injection and directed with external magnets to the right organ, which requires smaller dosage because of targeting, resulting in fewer side effects. Recently, Yu, et al. [41] reported a novel in vivo strategy for combined cancer imaging and therapy by employing thermally cross-linked superparamagnetic iron oxide nanoparticles as a drug-delivery carrier.

2.6.2   ENZYME-LINKED IMMUNOSORBENT ASSAYS (ELISA)

ELISA is a test used as a general screening tool for the detection of anti-bodies or antigens in a sample [42]. ELISA technology links a measurable enzyme to either an antigen or antibody. The procedure for detection of antibody (Ab) in patient's sample is as follows:

- Immobilize antigen (Ag) on the solid support (well);
- Incubate with patient sample;
- Add antibody-enzyme conjugate;
- Amount of antibody-enzyme conjugate bound is proportional to amount of Ab in the sample;
- Add substrate of enzyme;
- Amount of color is proportional to the amount of Ab in patient's sample.

The drawback of ELISA technique is that the procedure is time con-suming and it needs special equipment to run the assay (not portable). Thus, many techniques have been developed to fasten the process such as that of Xin, et al. [43] where they developed a chemiluminescence enzyme immunoassay using magnetic particles to monitor 17β-estradiol (E2) in environmental water samples. Another technique is using simple/rapid (S/R) test. The development of simple/rapid S/R tests has been extended from pregnancy detection of HIV antibodies in whole blood in addition to serum and plasma.

2.6.3   ANTIBIOTIC PRODUCTION

Penicillins are the most widely used β-lactam antibiotics. Production of antibiotics is one of the key areas in the field of applied microbiology. The conventional method of production is in stirred tank batch reactors. Because it is a no growth-associated process, it is difficult to produce the antibiotic in continuous fermentations with free cells. But it is a suitable case for cell immobilization, as growth and metabolic production can be uncoupled without affecting metabolite yields. Therefore, several attempts have been made to immobilize various microbial species on different sup-port matrices for antibiotic production. The most widely studied system is the production of penicillin G using immobilized cells of *Penicillium chrysogenum* [44]. In a recent study by Elnashar et al. [45] the authors were successful to covalently immobilize penicillin G acylase on carrageenan-modified gels with retention of 100% activity after 20 reuses.

2.6.4 MEDICAL APPLICATIONS

Medical applications of immobilized enzymes include diagnosis and treatment of diseases; these include enzyme replacement therapies, artificial cells and organs, and coating of artificial materials for better biocompatibility [46]. Examples of potential medical uses of immobilized enzyme systems are listed below:

- Asparaginase for leukemia;
- Arginase for cancer;
- Urease for artificial kidney, uremic disorders;
- Glucose oxidase for artificial pancreas;
- Carbonate dehydratase and catalase for artificial lungs;
- Glucoamylase for glycogen storage disease;
- Glucose-6-phosphate dehydrogenase for glucose-6-phosphate dehydrogenase deficiency;
- Xanthine oxidase for Lesch–Nyhan disease;
- Phenylalanine ammonia lyase for phenylketonuria;
- Urate oxidase for hyperuricemia;
- Heparinase for extracorporeal therapy procedures.

2.6.5 SOLVING THE PROBLEM OF LACTOSE INTOLERANCE

The consumption of foods with a high content of lactose is causing a medical problem for almost 70% of the world's population, especially in the developing countries, as the naturally present enzyme (β-galactosidase) in the human intestine loses its activity during lifetime [47]. Undigested lactose in chyme retains fluid, and bacterial fermentation of lactose results in production of gases, diarrhea, bloating, and abdominal cramps after consumption of milk and other dairy products. Unfortunately, there is no cure to "lactose intolerance." Interesting results of immobilized β-galactosidase on thermostable biopolymers of grafted carrageenan were obtained recently by Elnashar and Yassin [48, 49].

2.6.6   FRUCTOSE FOR DIABETICS AND FOR PEOPLE ON DIET REGIMEN

People on diet regimen and patients suffering from diabetes are highly rec-ommended to consume fructose rather than any other sugar. Fructose can be produced from starch by enzymatic methods, involving α-amylase, amy-loglucosidase, and glucose isomerase, resulting in the production of a mix-ture consisting of oligosaccharides (8%), fructose (45%), and glucose (50%). However, separation of fructose from this high content fructose syrup is costly and thus makes this method uneconomical. In industries, inulinases are used to produce 95% of pure fructose after one step of the enzymatic hydrolysis of inulin. Industrial inulin hydrolysis is carried out at 60°C to pre-vent microbial contamination and also because it permits the use of higher inulin substrate concentration due to increased solubility. Elnashar, et al., and Danial, et al. [48, 50] have succeeded recently to produce a thermostable inu-linolytic immobilized enzyme, which would be expected to play an important role in food and chemical industries, in which fructose syrup is widely used.

Superoxide dismutase (SOD) and catalase (CAT) have been encapsu-lated in biodegradable microspheres (MS) to obtain suitable sustained pro-tein delivery [48]. A modified water/oil/water double emulsion method was used for poly (D, L-lactide-co-glycolide) (PLGA) and poly (D, L-lactide) PLA MS preparation co-encapsulating mannitol, trehalose, and PEG400 for protein stabilization. SOD release from PLGA MS may be potentially useful for long-term sustained release of the enzyme for the treatment of rheumatoid arthritis or other intra-articular and joint diseases (inflamma-tory manifestation).

2.6.7   NONMEDICAL APPLICATIONS OF IMMOBILIZED ENZYMES

Treatment of pesticide-contaminated waste: Application of pesticide in agriculture serves to lower the cost of production, increase crop yields, provide better quality produce, and reduce soil erosion. Although pesticides are toxic and have adverse effect on human health and the environment, their use is inevitable in many cases as an effective means of controlling weeds, insect, fungus, parasites, and rodent pests. One of the most important technologies to be applied for this approach is immobilized enzyme. The immobilized enzyme is capable of breaking down a range of pesticide-contaminated waste, such as organophosphate insecticides [51, 52].

Neutralizing dangerous chemical gases or vapors: The use of immobilized enzymes in the national security arena has shown to be promising. For example, they could include infiltrating items such as air filters, masks, clothing, or bandages with the concentrated immobilized enzymes to neutralize dangerous chemical gases or vapors [53].

2.6.8 PURIFICATION OF PROTEINS

Protein purification is an important objective in industrial enzymes in order to increase the enzyme's specific activity and to obtain an enzyme in its pure form for a specific goal. Affinity ligands is the most used technique for the purification of target molecules as it can reduce the number of chromatographic steps in purification procedures to one or two steps. Immobilization of affinity ligands to an insoluble support can be a powerful tool in isolation of particular substances (e.g., protein) from a complex mixture of proteins. Some examples of affinity ligands are immobilized carbohydrate-binding proteins and immobilized metal ions. Another technique for protein purification is using electric field gradient focusing (EFGF) [54].

Immobilized carbohydrate-binding protein purification: proteins could be purified using immobilized carbohydrates such as mannose, lactose, and melibiose. For example, immobilized lactose on Sepharose 4B™ will be selective for purification of lactase from a mixture of other proteins.

EFGF: EFGF is a member of the family of equilibrium gradient focusing techniques (e.g., gel electrophoresis). It depends on an electric field gradient and a counter-flow to focus, concentrate, and separate charged analytes such as peptides and proteins. Because analytes with different electrophoretic mobilities have unique equilibrium positions, EFGF separates analytes according to their electrophoretic mobilities, similar to isoelectric focusing (IEF: electrophoresis is a pH gradient (where the cathode is at a higher pH value than the anode) that separates analytes according to isoelectric points. The constant counter flow is opposite to the electrophoretic force that drives the analytes. When the electrophoretic velocity of a particular analyte is equal and opposite to the velocity of the counter flow, the analyte is focused in a narrow band because at this position, the net force on it is zero. However, EFGF avoids protein precipitation that often occurs in IEF when proteins reach their isoelectric points and, therefore, can be applied to a broad range of proteins.

2.6.9   EXTRACTION OF BIOMOLECULES USING MAGNETIC PARTICLES

The traditional methods for biomolecule purification such as centrifugation, filtration, and chromatography can today be replaced by the use of magnetic particles. They are reactive supports for biomolecule capturing. Their use is simple, fast, and efficient for the extraction and purification of biomolecules. In the biomedical field, numerous publications deal with the use of magnetic particles for biomolecule extraction [55], cell sorting [56], and drug delivery [56]. Magnetic beads are widely used in molecular biology [58], medical diagnosis [59], and medical therapy [55]. The major application concerns the extraction of biomolecules such as proteins [61], antibodies, and nucleic acids [61]. Magnetic beads carrying antibodies are also used for specific bacteria [55] and virus capture [64].

2.6.9.1   Heavy Metal Removal

Heavy metal pollution is an environmental problem of worldwide concern. Several industrial wastewater streams may contain heavy metals such as Pb, Cr, Cd, Ni, Zn, As, Hg, Cu, and Ag. Traditionally, precipitation, solvent extraction, ion-exchange separation, and solid phase extraction are the most widely used techniques to eliminate the matrix interference and to concentrate the metal ions. Many materials have been used to remove them such as sorbents (e.g., silica, chitosan, sponge, etc.) and biosorbents (e.g., immobilized algae).

Biosorbents: can be defined as the selective sequestering of metal soluble species that result in the immobilization of the metals by microbial cells such as cyanobacteria. It is the physicochemical mechanisms of inactive (i.e., non-metabolic) metal uptake by microbial biomass. Metal sequestering by different parts of the cell can occur via various processes: complexation, chelation, coordination, ion exchange, precipitation, and reduction. Size of immobilized beads for metal removal is a crucial factor for use of immobilized biomass in the biosorption process. It is recommended that beads should be in the size range between 0.7 and 1.5 mm, corresponding to the size of commercial resins meant for removing metal ions.

2.6.10   PRODUCTION OF BIODIESEL

The idea of using biodiesel as a source of energy is not new [65], but it is now being taken seriously because of the escalating price of petroleum and, more significantly, the depletion of fossil fuels (oil and gas) within the next 35 years and the emerging concern about global warming that is associated with burning fossil fuels [66]. Biodiesel is much more environmentally friendly than burning fossil fuels, to the extent that governments may be moving toward making biofuels mandatory ("Biodiesel: Biodiesel Review," 2006. http://www.sipef. be/pdf/biodiesel_presentation.pdf). The global market survey of biodiesel has shown a tremendous increase in its production. Biodiesel is made by chemical combination of any natural oil or fat with an alcohol such as methanol and a catalyst (e.g., lipases) for the transesterification process. Transesterification is catalyzed by acids, alkalis, and lipase enzymes. Use of lipases offers important advantages as it is more efficient, highly selective, involves less energy consumption (reactions can be carried out in mild conditions), and produces less side products or waste (environmentally favorable). However, it is not currently feasible because of the relatively high cost of the catalyst [67]. The immobilized lipases most frequently used for biodiesel production are lipase B from *Candida antarctica* [68]. This is supplied by Novozymes under the commercial name Novozym 435® (previously called SP435) and is immobilized on an acrylic resin.

2.6.11   LIFE DETECTION AND PLANETARY EXPLORATION ANALYTICAL TECHNIQUES

Methods based on mass spectrometry have been traditionally used in space science. Planetary exploration requires the development of miniaturized apparatus for in situ life detection. Recently, a new approach is gaining acceptance in the space science community: the application of the well-known, highly specific, antibody–antigen affinity interaction for the detection and identification of organics and biochemical compounds. Antibody microarray technology allows scientists to look for the presence of thousands of different compounds in a single assay and in just one square centimeter. The detection of organic molecules of unambiguous biological origin is fundamental for the confirmation of present or past life. Preservation of biomarkers on the antibody stability under space environments, small molecule such as amino acids, purines, fatty acids can act as excellent biomarkers to search for line on Mars, despite being less resistant to oxidative degradation. Recent work by Kminek and Bada [69] showed that amino acids can be protected from radiolysis decomposition as long as they are shielded adequately from space radiation. They estimated that it is necessary to drill to a depth of 1.5 to 2 m to detect the amino acid signature of life that became extinct about three billion years ago.

Recently, antibody microarray, a new immobilization technology that kept the stability of antibody under space environment allowed it to be applied for planetary exploration Exomars mission ("Exomars Mission Conference," 2005. http://www.aurora.rl.ac.uk/Report_of_Pasteur_9_Sept.pdf).

2.7   NEW TECHNOLOGIES FOR ENZYME IMMOBILIZATION

Single Enzyme Nanoparticles: In the field of industrial enzymes, extensive research is being conducted for improving the enzyme stability under harsh conditions. As an innovative way of enzyme stabilization, "single-enzyme nanoparticles (SENs)" technology was rather attractive because enzymes in the nanoparticle exhibited very good stability under harsh conditions [70]; the authors developed armored SENs that surround each enzyme molecule with a porous composite organic/inorganic network of less than a few nanometers thick. They significantly stabilized chymotrypsin and trypsin and the protective covering around chymotrypsin is so thin and porous that a large mass transfer limitation on the substrate could not take place. Yan, et al. [71] provided a simple method that yields a single enzyme capsule with enhanced stability, high activity, and uniform size. The 2-step procedure includes surface acryloylation and in situ aqueous polymerization to encapsulate a single enzyme in nanogel to provide robust enzymes for industrial biocatalysis.

The immobilized horseradish peroxidase (HRP) exhibited similar biocatalytic behavior (Km and $k$cat) to the free enzyme. However, the immobilization process significantly improved the enzyme's stability at high temperature in the presence of polar organic solvent.

Enzymatic Immobilization of Enzyme: The use of green chemistry rather than using harsh chemicals is one of the main goals in enzyme industries to avoid the partial denaturation of enzyme protein. An emerging and novel technology is to fabricate solid protein formulations. As model proteins, enhanced green fluorescent protein (EGFP) and glutathione S-transferase (GST) were tagged with a neutral Gln-donor substrate peptide for MTG (Leu-Leu-Gln-Gly, LLQG-tag) at their C-terminus and immobilized onto the casein-coated polystyrene surface [108]. Luciferase (Luc) and GST ybbR-fusion proteins were immobilized onto PEGA resin retaining high levels of enzyme activity using phosphopantetheinyl transferase (Sfp) mediating site-specific covalent immobilization [72].

In general, the Sfp-catalyzed surface ligation is mild, quantitative, and rapid, occurring in a single step without prior chemical modification of the target protein.

Microwave Irradiation: The use of porous supports for immobilization of enzymes is difficult to distribute because of diffusion limitations [60]. For enzymes with large dimensions such as penicillin acylase (PA), the mass transfer is even slower. The immobilization of such enzyme to porous materials can prove tedious using conventional techniques. The activities of papain and penicillin acylase immobilized with microwave-assisted method were 779.6 and 141.8 U/mg, respectively. In another experiment, macromolecules crowding was combined with small molecular quenching to perfect microwave-assisted covalent immobilization [63].

2.7.1   PHOTOIMMOBILIZATION TECHNOLOGY

In the field of immobilization of biomolecules, potential applications of photoimmobilization using nitrene groups could be used. Nitrene groups have a property of insertion into the C-H bond. When photoreactive polymer and horseradish peroxidase or glucose oxidase are exposed to ultraviolet (UV) light at 365 nm, the reactive nitrene immobilizes the protein molecules in 10 to 20 min through covalent bonding [28].

2.7.1.1   Horseradish Peroxidase (HRP) and Glucose Oxidase (GOD)

These enzymes have been immobilized onto the photoreactive cellulose membrane by ultraviolet radiation and sunlight [57]. The authors found that sunlight intensity required for optimum immobilization was 21,625 lux, beyond which no appreciable increase in immobilization was observed. Moreover, sunlight exposure gave better immobilization than 365 nm UV light.

Ionic liquids: Ionic liquids, the green solvents for the future, are composed entirely of ions, and they are salts in the liquid state. In the patent and academic literature, the term "ionic liquid" now refers to liquids composed entirely of ions that are fluid around or below 100°C (e.g., ethanolamine nitrate, m.p. 52–55°C). The date of discovery of the "first" ionic liquid is disputed, along with the identity of the discoverer.

Room-temperature ionic liquids are frequently colorless, fluid, and easy to handle [61]. Versatile biphasic systems could be formed by controlling the aqueous miscibility of ionic liquid [68]. Based on a biphasic catalytic system where the enzyme is immobilized into an ionic liquid (IL), Mecerreyes and co-workers [66] have reported a new method that allows recycling and re-using of the HRP enzyme in the biocatalytic synthesis of PANI is polyaniline (PANI). The HRP enzyme was dissolved into the IL

1-butyl-3-methylimidazolium hexafluorophosphate and the IL/HRP phase acts as an efficient biocatalyst and can be easily recycled and reused several times. Due to the immiscibility between the IL and water, the immobilized HRP could be simply recovered by liquid/liquid phase separation after the biocatalytic reaction [64, 34]. Although this new method is faster and easier than the classical immobilization of HRP into solid supports, it would not be widely applied to the industrial production in the coming future because of the ionic liquids' cost.

2.7.1.2   Recommendation for the Future of Immobilization Technology

At present, a vast number of methods of immobilization are currently available. Unfortunately, there is no universal enzyme support, i.e., the best method of immobilization might differ from enzyme to enzyme, from application to application, and from carrier to carrier. Accordingly, the approaches currently used to design robust industrial immobilized enzymes are, without exception, labeled as "irrational," because they often result from screening of several immobilized enzymes and are not designed. As a consequence, some of the industrial enzymes are working below their optimum conditions. Recently, Cao [5] in his book "Carrier bound immobilized enzymes" tackled this problem as he surmised that the major problem in enzyme immobilization is not only the selection of the right carrier for enzyme immobilization but it is also how to design the performance of the immobilized enzyme. The author of this review article has been working in this field for the last 10 years and suggests to follow these steps (from his point of view) in order to achieve this goal in the shortest time:

1. build a database containing all information on the available biomolecules (enzymes, antibodies, etc.) and carriers (organic, inorganic, magnetic hybrid, ionic liquids, etc.),
2. use the dry lab (bioinformatics) to validate the probability of success and the efficiency of the immobilization process, and
3. start the experiment in the wet lab.

The author believes that if this strategy could be implemented, we should expect immobilized molecules working at their optimum conditions, with higher stability and efficiency, which will save money, time, and effort for the prosperity of humans.

2.8   IMMOBILIZATION OF CELLS

Immobilization can be done using the method of choice of an individual enzyme or the whole cell depending on application. Many processes such as enhancement of cleaning potency of a detergent can be achieved by incorporation of immobilized lipases or proteases in the preparation. This needs a single enzyme to be immobilized. Many enzyme catalyzed process are not hindered by the presence of other enzymes. For such processes immobilized whole cell can be used which is a repository of numerous enzymes. Whole cell immobilization is cost effective as compared to single enzyme immobilization. Processes such as sewage treatment needs a number of enzymes, and for this purpose, whole cells are immobilized, as they serve as a source of multienzyme systems. Thus, immobilized cells have been traditionally used for the treatment of sewage. The commonly employed technique for whole cell immobilization are entrapment and surface attachment. Immobilized whole cells can be viable or nonviable.

2.8.1   IMMOBILIZATION OF VIABLE CELLS

Certain fermentation processes need the presence of viable cells; therefore, to preserve the viability during immobilization, mild procedures are employed. Cells grown in culture can act as immobilized cells that are viable.

2.8.2   IMMOBILIZATION OF NONVIABLE CELLS

Viability of cells is not a prerequisite for many industrial fermentation procedures; thus, the whole cell can be immobilized and subjected to harsh treatment, because even if the cell dies, the immobilized part would have the enzyme of choice and cause reaction. In many instances, immobilized nonviable cells are preferred over the enzymes or even the viable cells due to costly isolation and purification processes. Glucose isomerase is an example of nonviable cell immobilization for the industrial production of high fructose syrup. Glucose isomerase is an intracellular enzyme produced by a number of microorganisms. The species of *Arthrobacter, Bacillus,* and *Streptomyces* are the preferred sources.

Other important examples of microbial biocatalysts and their applications are given in Table 2.2.

2.8.3 IMMOBILIZED ENZYMES IN BIOCHEMICAL ANALYSIS

Another application of immobilized enzymes (or cells) includes the development of precise and specific analytical techniques for the estimation of several biochemical compounds. Enzyme acts on substrate to form products. Thus, the analytical biochemical process utilizes this property where a decrease in the substrate concentration or an increase in the product level or an alteration in the cofactor concentration can be used as an indicator for the assay.

2.8.4 IN AFFINITY CHROMATOGRAPHY AND PURIFICATION

Immobilized enzymes can be used in affinity chromatography. Based on the property of affinity, it is possible to purify several compounds, e.g., antigens, antibodies, and cofactors.

TABLE 2.2 Immobilized Cells Used Industrially

Immobilized microorganism	Application (s)
E. coli	L-aspartic acid synthesis from fumaric acid and NH3
E. coli	L-tryptophan synthesis from serine and indole
Pseudomonas spp	L-serine synthesis from glycine and methanol
Saccharomyces cerevisiae	Sucrose hydrolysis; large scale alcohol production
Zymomonas mobilis	Sorbitol and gluconic acid synthesis from glucose and fructose
Humicola sp	Rifamycin B conversion to Rifamycin S
Pseudomonas chlororaphis	Acrylamide production from acrylonitrile
Several bacteria and yeasts	Biosensors

2.9   SUMMARY

The immobilization of enzymes is a useful tool to meet cost targets and has a number of technological advantages, for instance, it enables repeated use of enzyme and hence produces significant cost savings. Moreover, immobilized enzyme can easily be separated from the reaction liquid and thereby reduce laborious separation steps. Additional benefits arise from stabilization against harsh reaction condition, which are deleterious to soluble enzyme preparation. Cross-linked enzyme crystals (CLECs) is a method of immobilizing enzyme, i.e., converting soluble form to insoluble counterpart. CLEC is a new technique and very promising type of biocatalyst for industrial application. Moreover, new nanomaterials will be produced by engineering specific covalent linkages into protein or enzyme crystals. Strategic modifications in the surface of the protein via site-directed mutagenesis, followed by cross-linking of the crystals of these mutants along the crystallographic planes will produce sheets or fibers upon subsequent dissociation of the crystal. It is intended to exploit this new class of nanomaterials for application such as ultrafiltration membrane, enzyme crystal morphologies, and ultimately microelectronic devices. It promises the stability of three-dimensional cross-linked enzyme crystals without the same substrate and product mass transfer resistance.

2.10   REVIEW QUESTIONS

1. What is the importance of enzyme immobilization?
2. What are the different techniques for enzyme immobilization?
3. Discuss entrapment and covalent binding in relation to enzyme immobilization.
4. What are the industrial applications of immobilized enzyme?
5. How can immobilized enzymes play a critical role in point-of-care patient management?
6. How can immobilized enzymes be used to ease the burden of environmental pollution?

KEYWORDS

- **entrapment**
- **enzyme immobilization**
- **immobilization by covalent bonding**
- **medical application enzyme immobilization**

REFERENCES

1. Carpentier, B., & Cerf, O., (1993). Biofilms and their consequences, with particular reference to hygiene in the food industry. *J. Appl. Bacteriol.*, *75*(6), 499–511.
2. Danial, E. N., Elnashar, M. M., & Awad, G. E., (2010). Immobilized inulinase on grafted alginate beads prepared by the one-step and the two-steps methods indus. *Chem. Eng. Res.*, *49*(7), 3120–3125.
3. Nelson, J. M., & Griffin, E. G., (1916). Adsorption of invertase. *J. Am. Chem. Soc.*, *38*, 1109–1115.
4. Tosa, T., Mori, T., Fuse, N., & Chibata, I., (1967). Studies on continuous enzyme reactions part V kinetics and industrial application of aminoacylase column for continuous optical resolution of Acyl-Dl amino acids. *Biotechnol. Bioeng.*, *9*(4), 603–615.
5. Menaa, B., Torres, C., Herrero, M., Rives, V., Gilbert, A. R. W., & Eggers, D. K., (2008). Protein adsorption to organically-modified silica glass leads to a different structure than sol-gel encapsulation. *Biophys. J.*, *95*(8), 51–53.
6. Cetinus, S., Sahin, E., & Saraydin, D., (2009). Preparation of Cu (II) adsorbed chitosan beads for catalase immobilization. *Food Chem.*, *114*(3), 962–969.
7. Cao, L., & Schmid, R. D., (2005). *Carrier-Bound Immobilized Enzymes: Principles, Application and Design*, WILEY-VCH Verlag GmbH & Co: Weinheim, 132–134.
8. Menaa, B., Miyagawa, Y., Takahashi, M., Herrero, M., Rives, V., Menaa, F., & Eggers, D. K., (2009). Bioencapsulation of apomyoglobin in nanoporous organosilica sol-gel glasses: Influence of the siloxane network on the conformation and stability of a model protein. *Biopolymers*, *91*(11), 895–906.
9. Patil, J. S., Kamalapur, M. V., Marapur, S. C., & Kadam, D. V., (2010). Ionotropic Gelation and polyelectrolyte complexation: The novel techniques to design hydrogel particulate sustained, modulated drug delivery system: A review. *Dig. J. Nanomater. Biostruct.*, *5*, 241.
10. Xie, T., Wang, A., Huang, L., Li, H., Chen, Z., Wang, Q., & Yin, X., (2009). Review: recent advance in the support and technology used in enzyme immobilization. *Afr. J. Biotechnol.*, *8*(19), 4724–4733.

11. Chen, B., Miller, M. E., & Gross, R. A., (2007). Effects of porous polystyrene resin parameters on candida antarctica lipase B adsorption, distribution, and polyester synthesis activity. *Langmuir.*, *23*(11), 6467–6474.

12. Kim, M. I., Kim, J., Lee, J., Jia, H., Bin Na, H., Youn, J. K., Kwak, J. H., Dohnalkova, A., Grate, J. W., & Wang, P., (2007). Cross-linked enzyme aggregates in hierarchically-ordered mesoporous Silica: A simple and effective method for enzyme stabilization, *Biotechnol. Bioeng.*, *96*(2), 210–218.

13. Wang, A. M., Zhou, C., Wang, H., Shen, S. B., Xue, J. Y., & Ouyang, P. K., (2007). Covalent assembly of penicillin acylase in mesoporous silica based on macromolecular crowding theory. *Chinese J. Chem. Eng.*, *15*(6), 788–790.

14. Wang, A., Wang, H., Zhu, S., Zhou, C., Du, Z., & Shen, S., (2008). An efficient immobilizing technique of penicillin acylase with combining mesocellular silica foams support and P-benzoquinone cross linker. *Bioprocess Biosyst. Eng.*, *31*(5), 509–517.

15. Correa-Basurto, J. I., Vazquez-Alcantara, E., Terres-Rojas., & Trujillo-Ferrara, J., (2006). Catalytic activity of acetylcholinesterase immobilized on mesoporous molecular sieves, *Int. J. Biol. Macromolec.*, *40*(5), 444–448.

16. Boller, T., Meier, C., & Menzler, S., (2002). Eupergit oxirane acrylic beads: How to make enzymes fit for biocatalysis. *Org. Process Res. Dev.*, *6*, 509–519.

17. Chong, A. S. M., & Zhao, X. S., (2004). Design of large-pore mesoporous materials for immobilization of penicillin G acylase biocatalyst. *Catal. Today*, *93–95*, 293–299.

18. Lei, C. H., Shin, Y. S., Liu, J., & Ackerman, E. J., (2002). Entrapping enzyme in a functionalized nanoporous support. *J. Am. Chem. Soc.*, *124*(38), 11242–11243.

19. Wang, A. M., Liu, M. Q., Wang, H., Zhou, C., Du, Z. Q., Zhu, S. M., Shen, S. B., & Ouyang, P. K., (2008). Improving enzyme immobilization in mesocellular siliceous foams by microwave irradiation. *J. Biosci. Bioeng.*, *106*(3), 286–291.

20. Cheung, M. S., & Thirumalai, D., (2006). Nanopore-protein interactions dramatically alter stability and yield of the native state in restricted spaces, *J. Mol. Biol.*, *357*(2), 632–643.

21. Szamocki, R., Velichko, A., Mucklich, F., Reculusa, S., Ravaine, S., Neugebauer, S., Schuhmann, W., Hempelmann, R., & Kuhn, A., (2007). Improved enzyme immobilization for enhanced bioelectrocatalytic activity of porous electrodes. *Electrochem. Commun.*, *9*(8), 2121–2127.

22. Chen, R. J., Zhang, Y. G., Wang, D. W., & Dai, H. J., (2001). Noncovalent sidewall functionalization of single- walled carbon nanotubes for protein immobilization. *J. Am. Chem. Soc.*, *123*(16), 3838–3839.

23. Wan, L. S., Ke, B. B., & Xu, Z. K., (2008). Electrospun nanofibrous membranes filled with carbon nanotubes for redox enzyme immobilization. *Enzyme Microb. Technol.*, *42*(4), 332–339.

24. Bayramoglu, G., Kiralp, S., Yilmaz, M., Toppare, L., & Arica, M. Y., (2008). Covalent immobilization of chloroperoxidase onto magnetic beads: Catalytic properties and stability. *Biochem. Eng. J.*, *38*(2), 180–188.

25. Dyal, A., Loos, K., Noto, M., Chang, S. W., Spagnoli, C., Shafi, K., Ulman, A., Cowman, M., & Gross, R. A., (2003). Activity of candida rugosa lipase immobilized on gamma-Fe2o3 magnetic nanoparticles. *J. Am. Chem. Soc.*, *125*(7), 1684–1685.

26. Liu, X. Q., Guan, Y. P., Shen, R., & Liu, H. Z., (2005). Immobilization of lipase onto micron-size magnetic beads. *J. Chromatogr. B-Analyt. Technol. Biomed. Life Sci.*, *822*(1–2), 91–97.

27. Sadasivan, S., & Sukhorukov, G. B., (2006). Fabrication of hollow multifunctional spheres containing MCM-41 nanoparticles and magnetite nanoparticles using layer-by-layer method. *J. Colloid Interface Sci.*, *304*(2), 437–441.

28. Naqvi, A., & Nahar, P., (2004). "Photochemical immobilization of proteins on microwave-synthesized photoreactive polymers," *Analytical Biochemistry*, *327*(1), pp. 68–73.

29. Feng, W., & Ji, P., (2011). Enzymes immobilized on carbon nanotubes. *Biotechnol. Adv.*, *29*, 889–895.

30. Gupta, M. N., Kaloti, M., Kapoor, M., & Solanki, K., (2011). Nanomaterials as matrices for enzyme immobilization. *Artif. Cells Blood Substit. Immobil. Biotechnol.*, *39*, 98–109.

31. Ansari, S. A., & Husain, Q., (2012). Potential applications of enzymes immobilized on/in nano materials: *A* review. *Biotechnol. Adv.*, *30*, 512–523.

32. Verma, M. L., Barrow, C. J., & Puri, M., (2013). Nanobiotechnology as a novel paradigm for enzyme immobilisation and stabilisation with potential applications in biodiesel production. *Appl. Microbiol. Biotechnol.*, *97*, 23–39.

33. Hwang, E. T., & Gu, M. B., (2013). Enzyme stabilization by nano/microsized hybrid materials. *Eng. Life Sci.*, *13*, 49–61.

34. Van Rantwijk, F., Lau, R. M., & Sheldon, R. A., (2003). "Biocatalytic transformations in ionic liquids," *Trends Biotechnology*, *21*(3), pp. 131–138.

35. Vertegel, A. A., Siegel, R. W., & Dordick, J. S., (2004). Silica nanoparticle size influences the structure and enzymatic activity of adsorbed lysozyme. *Langmuir.*, *20*, 6800–6807.

36. Lan, D., Li, B., & Zhang, Z., (2008). Chemiluminescence flow biosensor for glucose based on gold nanoparticle-enhanced activities of glucose oxidase and horseradish peroxidase. *Biosens. Bioelectron.*, *24*, 940–944.

37. Wu, C. L., Chen, Y. P., Yang, J. C., Lo, H. F., & Lin, L. L., (2008). Characterization of lysine-tagged Bacillus stearothermophilus leucine aminopeptidase II immobilized onto carboxylated gold nanoparticles. *J. Mol. Catal. B: Enzym.*, *54*, 83–89.

38. Keighron, J. D., & Keatin, C. D., (2010). Enzyme: nanoparticle bioconjugates with two sequential enzymes: stoichiometry and activity of malate dehydrogenase and citrate synthase on Au nanoparticles. *Langmuir.*, *26*, 18992–19000.

39. Pack, D. W., Hoffman, A. S., Pun, S., & Stayton, P. S., (2005). Design and development of polymers for gene delivery. *Nat. Rev. Drug Discov.*, *4*, 581–593.

40. Grayson, A. C. R., Choi, I. S., Tyler, B. M., Wang, P. P., & Michael, B. H., (2003). Multi-pulse drug delivery from a resorbable polymeric microchip device. *Nature Materials J. Cima.*, *2*(11), 767–772.

41. Yu, M., Jeong, Y., Park, J., Park, S., Kim, J., Min, J., Kim, K., & Jon, S., (2008). Drug-loaded superparamagnetic iron oxide nanoparticles for combined cancer imaging and therapy in vivo. *Angewandte Chem. Int. Ed.*, *47*(29), 5362–5365.

42. Farre, M., Kuster, M., Brix, R., Rubio, F., Alda, M. J. L., & Barcelo, D., (2007). Comparative study of an estradiol enzyme-linked immunosorbent assay kit, liquid chromatography-tandem mass spectrometry, and ultra performance liquid chroma-

tography-quadrupole time of flight mass spectrometry for part-per-trillion analysis of estrogens in water samples. *J. Chromatogr. A., 1160*(1–2), 166–175.

43. Xin, T., Wang, X., Jin, H., Liang, S., Lin, J., & Li, Z., (2009). Development of magnetic particle-based chemiluminescence enzyme immunoassay for the detection of 17β-estradiol in environmental water. *Appl. Biochem. Biotechnol., 158*(3), 582–594.

44. Carpentier, B., & Cerf, O., (1993). Biofilms and their consequences, with particular reference to hygiene in the food industry. *J. Appl. Bacteriol., 75*(6), 499–511.

45. Elnashar, M. M., Yassin, A. M., & Kahil, T., (2008). Novel thermally and mechanically stable hydrogel for enzyme immobilization of penicillin G acylase via covalent technique. *J. Appl. Polym. Science, 109*(6), 4105–4111.

46. Piskin, A. K., (1993). Therapeutic potential of immobilized enzymes. *NATO ASI Series, Series E., 252*, 191.

47. Richmond, M., Gray, J., & Stine, C., (1981). Beta-galactosidase: Review of recent research related to technological application, nutritional concerns, and immobilization, *The Journal of Dairy Science, 1759*, 64.

48. Elnashar, M. M., Danial, E. N., & Awad, G. E., (2009). Novel carrier of grafted alginate for covalent immobilization of inulinase. *Industrial & Engineering Chemistry Research, 48*(22), 9781–9785.

49. Elnashar, M. M., & Yassin, A. M., (2009). Lactose hydrolysis by β-galactosidase covalently immobilized to thermally stable biopolymers, *Applied Biochemistry and Biotechnology, 159*(2), 426–437.

50. Danial, E. N., Elnashar, M. M., & Awad, G. E., (2010). Immobilized inulinase on grafted alginate beads prepared by the one-step and the two-steps methods indus, *Chemical Engineering Research, 49*(7), 3120–3125.

51. Horne, I., Sutherland, T. D., Harcourt, R. L., Russell, R. J., & Oakeshott, J. G., (2002). Identification of an (Organophosphate degradation) gene in an agrobacterium isolate, *Applied and Environmental Microbiology, 68*(7), 3371–3376.

52. Sharmin, F., Rakshit, S., & Jayasuriya, H., (2007). Enzyme immobilization on glass surfaces for the development of phosphate detection biosensors, *Agricultural Engineering International: The CIGR Ejournal.* Manuscript FP 06 019, Vol. IX.

53. Ackerman, E., & Lei, C., (2008). *"Immobilizing Enzymes for Useful Service."* http//www.google.19.11.2008.

54. *Handbook: Purifying Challenging Proteins: Principles and Methods"* in 2007, Hand book from GE Healthcare, "Purifying Challenging Proteins: Principles and Methods," General Electric Co, USA.

55. Delair, T., & Meunier, F., (1999). Amino-containing cationic latex oligo-conjugates: Application to diagnostic test sensitivity enhancement, *Colloids and Surface, 153*(1–3), 341–353.

56. Kemshead, J. T., Treleaven, J. G., Gibson, F. M., Ugallstad, J., Rembaum, A., & Philip, T., (1985). Removal of malignant cells from marrow using magnetic microspheres and monoclonal antibodies, *Progress in Experimental Tumor Research, 29*, 249–245.

57. Kumar, S., & Nahar, P., (2007). "Sunlight-induced covalent immobilization of proteins," *Talanta, 71*(3), pp. 1438–1440.

58. Andreadis, J. D., & Chrisey, L. A., (2000). Use of immobilized PCR primer to generate covalently immobilized DNAs for in vitro transcription/translation reaction, *Nucleic Acids Research, 28*(2), e5.

59. Myrmel, M., Rimstad, E., & Wasteson, Y., (2000). IMS of norwalk-like virus (Geno group I) in artificially contaminated environmental water samples, *International Journal of Food Microbiology, 62*(1–2), 17–26.

60. Buchholz, K., (1979). "Non-uniform enzyme distribution inside carriers," *Biotechnology Letters, 1*(11), pp. 451–456.

61. Ding, X., & Jiang, Y., (2000). Adsorption/desorption of protein on magnetic particles covered by thermosensitive polymers, *Journal of Applied Polymer Science, 278*, 459.

62. Rogers, R. D., & Seddon, K. R., (2003). "Ionic liquids-solvents of the future?" *Science, 302*(5646), pp. 792–793.

63. Rolf Scharer, Md. Hossain, M., & Do, D. D., (1992). Determination of total and active immobilized enzyme distribution in porous solid supports. *Biotechnology and Bioengineering, 39*(6), 679–687.

64. Sheldon, R. A., Lau, R. M., Sorgedrager, M. J., Van Rantwijk, F., & Seddon, K. R., (2002). "Biocatalysis in ionic liquids," *Green Chemistry, 4*, pp. 147–151.

65. Sawayama, S., Inoue, S., Dote, Y., & Yokoyama, S. Y., (1995). CO_2 fixation and oil production through microalga, *Energy Conversion and Management, 36*(6–9), 729–731.

66. Rumbau, V., Marcilla, R., Ochoteco, E., Pomposo, J. A., & Mecerreyes, D., (2006). "Ionic liquid immobilized enzyme for biocatalytic synthesis of conducting polyaniline," *Macromolecules, 39*(25), pp. 8547–8549.

67. Fukuda, H., Kondo, A., & Noda, H., (2001). Biodiesel fuel production by transesterification of oils, *Journal of Bioscience and Bioengineering, 92*(5), 405–416.

68. Gutowski, K. E., Broker, G. A., Willauer, H. D., Huddleston, J. G., Swatloski, R. P., Holbrey, J. D., & Rogers, R. D., (2003). "Controlling the aqueous miscibility of ionic liquids: Aqueous biphasic systems of water-miscible ionic liquids and water-structuring salts for recycle, metathesis, and separations," *Journal of American Chemical Society, 125*(22), pp. 6632–6633.

69. Kminek, G., & Bada, J. L., (2006). The effect of ionizing radiation on the preservation of amino acids on mars. *Earth Planet, Science Letters, 245*(1–2), 1–5.

70. Kim, J., & Grate, J. W., (2003). Single-enzyme nanoparticles armored by a nanometer-scale organic/inorganic network, *Nano Letters, 3*(9), 1219–1222.

71. Yan, M., Ge, J., Liu, Z., & Ouyang, P. K., (2006). "Encapsulation of single enzyme in nanogel with enhanced biocatalytic activity and stability," *Journal of American Chemical Society, 128*(34), pp. 11008–11009.

72. Wong, L. S., Thirlway, J., & Micklefield, J., (2008). "Direct site-selective covalent protein immobilization catalyzed by a phosphopantetheinyl transferase," *Journal of American Chemical Society, 130*(37), pp. 12456–12464.

CHAPTER 3

ENZYMES IN NON-CONVENTIONAL MEDIA

ANJALI PRIYADARSHINI and PRERNA PANDEY

CONTENTS

3.1   INTRODUCTION

The environment inside the cells is mixed, with hydrophilic in some parts and hydrophobic in other parts of the cell. Thus, enzymatic reactions outside the cell have been performed in buffered aqueous phase. One of the most challenging subjects of study in applied biocatalysis is the use of reaction media that are significantly less polar than water. The interest in the use of such nonconventional media has grown immensely after the recognition that biocatalysts such as enzymes, cell organelles and whole cells are active in all kinds of nonconventional media such as organic solvents, gases, and even supercritical fluids. The application of biocatalysts in these nonconventional media is very attractive in organic synthesis routes that desire the advantageous characteristics of biocatalysts such as high substrate specificity and that involve substrates and products that are poorly soluble in water. When water is one of the products of the synthesis reaction, the application of these media is even more attractive, because the equilibrium may be shifted in favor of the synthesis by lowering the water activity or by altering the partitioning behavior of the substrates and products between the relevant phases. Other reasons to justify the use of nonconventional media are [1]:

- Increased solubility of nonpolar substrates and products.
- Shift in thermodynamic equilibria in favor of synthetic reactions.
- Reduction in water activity leads to reduction of water-dependent side reactions.
- Reduction of substrate and/or product inhibition.
- Enables manipulation of the stereo and regioselectivity of an enzymic biocatalyst.

In the pharmaceutical industry, a common practice is to use nonconventional media such as nonaqueous heterogeneous systems for the production of flavors compounds, oleochemicals, and drug intermediates.

Organic solvents have also been used for steroid bioconversion. This has become the method of choice because of low solubility of substrate and product steroids in water; hence, they can be solubilized using an organic solvent system.

This chapter deals with some of the current trends in biocatalysis in systems, mainly focusing on organic solvent systems, reverse micelles, and supercritical fluids along with applications of ionic liquids (ILs) as "green" solvents in:

- biocatalytic transformations of commercially important compounds.
- extractions of a variety of substances, including metal ions, organic and biomolecules, and organosulfur from fuels and gases.

3.2   POTENTIAL OF BIOCATALYSIS IN ORGANIC SOLVENTS

There are several potential advantages for the introduction of organic solvents in synthetic reactions. Organic solvents will increase the solubility of poorly water-soluble substrates, thereby improving the volumetric productivity of the reaction. The thermodynamic reaction equilibrium may be shifted to favor synthesis over hydrolysis, either by altering the partitioning of the substrate/product between the phases of interest or by reducing the water activity. The latter can be achieved by replacing the water in the reaction mixture by a water-miscible organic solvent or by introduction of polymers, sugars, or salts.

Reduction of the water activity or the associated water content will also diminish water-dependent unwanted side reactions such as polymerization of oxidized phenols [2] or hydrolysis during transesterification reactions [3]. In addition, higher product yields will be achieved by reduction of substrate and/or product inhibition, either indirectly by maintaining a low concentration in the aqueous microenvironment of the biocatalyst [4, 5] or directly by changing the interactions between the inhibitor and the active site of the enzyme [6].

Application of low-boiling organic solvents will simplify recovery of the product and biocatalyst. The biocatalyst does not dissolve in the solvent and can thus easily be recovered from the reaction mixture, for instance by filtration, while the product can be obtained by evaporating the solvent, provided there is sufficient difference in boiling point. Other advantages of enzymatic catalysis in organic solvents are improved thermostability of the enzyme, particularly when microaqueous reaction media are used [7, 8] and the possibility to manipulate the stereo- and regio-selectivity of the enzyme in such media [9, 10].

Obviously, not all these advantages are relevant to all the categories of organic solvent reaction media and reactions, and of course, disadvantages of using organic solvents in biocatalysis also exist, e.g., the organic solvent may denature or inhibit the biocatalyst. In addition introduction of organic solvent into reaction mixture facilitates recovery of insoluble enzymes.

3.3   ORGANIC SOLVENT REACTION MEDIA

Four categories of organic solvent reaction media for biocatalysis can be distinguished (Figure 3.1). The water/organic solvent mixtures may consist mainly of water with a relatively small amount of a water-miscible solvent (A). The mixture may consist of a two-phase system of a water-immiscible organic solvent and an aqueous buffer (Bl, B2, and B3) or it may be an organic solvent in which dry biocatalyst is suspended, so-called microaqueous organic solvent mixtures (C). In the latter case, the water present is located mostly on the solid enzyme particles. The fourth category of organic solvent reaction media is the reversed micelles (D). Reversed micelles consist of tiny droplets of aqueous medium (radii in the range of 1–50 nm) stabilized by surfactant in a bulk of water

3.4   SUPERCRITICAL FLUIDS

Promising types of nonconventional medium for biocatalysis are the super- and near-critical fluids. Supercritical fluids (SCFs) are those compounds that exist at a temperature and a pressure above their corresponding critical value. Supercritical fluids are highly compressed gases that combine properties of gases and liquids in an interesting

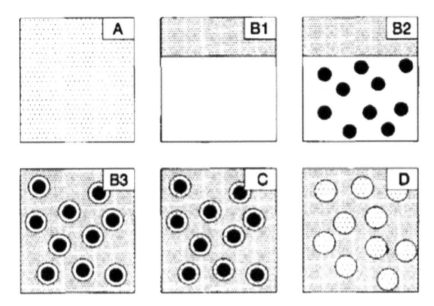

FIGURE 3.1 Schematic representation of the four categories of organic solvent reaction media. A: water-miscible solvent; Bl: two-phase system, low volume organic solvent, solubilized biocatalyst; B2: two-phase system, low volume organic solvent, immobilized biocatalyst; B3: two-phase system, aw = 1, high volume organic solvent, immobilized biocatalyst; C: micro-aqueous system, aw < 1; D: reversed micelles

manner. Examples of such SCFs include supercritical xenon, ethane, and carbon dioxide with a very wide application in analytical and synthetic chemistry.

These SCFs have the possibility of wide range of varying solvent power. This property can be explained by the fact that they can be used as good solvents when operating conditions leads to a high specific gravity as during high pressure or temperature that is near the critical temperature. This solvent nature can be utilized as in or as extraction solvents, chromatography eluents, and reaction media. Apart from this, they can also be converted into compressed gas with low solvent when conditions are pressure below the critical pressure and temperature over the liquefaction temperature at this pressure. This property can be used to perform fluid-solute separation, thus exemplifying the range in which it can be used as a very good to poor solvent, depending on its application.

The use of enzymes in nonconventional media such as SCFs helps in improving the activity and utility of such enzymes in anhydrous environments. Common examples of SCFs are mainly carbon dioxide, freons, hydrocarbons, or inorganic compounds such as SF6, N_2O, etc. The most commonly used system is supercritical carbon dioxide ($SCCO_2$). This is due to the fact that its critical point of 73.8°C and 31.1°C making the equipment design and reaction set-up relatively simple as compared to other SCFs. The bulk transport of substrate to enzyme is greatly increased by the property of high diffusivities and low viscosities in SCFs. Solubilization of substrates is generally better in SCFs than that in organic liquid solvents owing to the fact that high level of diffusivity and low viscosity increase the transfer of substrate to enzyme, thus enhancing the enzyme-catalyzed reaction. Purity of any product increases when there is no organic solvent which at times is difficult to separate from the end product, which is very desirable in any food product.

Some loss in the enzyme activity can occur in these systems during the depressurization of the reaction mixture, leading to a mixed opinion on the economic viability of such systems compared to organic liquid systems as in ester synthesis reactions in $SCCO_2$ [11]. A small change in pressure or temperature near the critical point may result in great change in viscosity of SCF, thus leading to change in diffusivity and solubility of compounds dissolved in it. In enzymatic reactions, interaction between carbon dioxide and enzyme molecules led to conformational changes in the enzymes, which can cause pre-active sites to emerge to catalyze a reaction in the critical region. The high diffusion rate can also facilitate product separation. In addition, the use of s $SCCO_2$ can have adverse effects on enzymes by decreasing the pH of the microenvironment of the enzyme and by the formation of carbamates due to covalent modification of free amino groups at the surface of the protein [11, 12]. Therefore, $SCCO_2$ medium in the near critical region ideally leads to enzyme activation by causing movement of its surface groups and creating active sites [13]. The properties of carbon dioxide, i.e. nontoxic, chemically inert, and ease of separation after the reaction, form the crux of its use as SCFs. Further, the very high diffusibility of SCFs results in increase in transport and thus enhances the rate of bioconversion.

However, carbon dioxide always behaves as a "non-polar" solvent that selectively dissolves the lipids that are water-insoluble compounds, including vegetable oils, butter, fats, hydrocarbons, etc., but not hydrophilic compounds like sugars and proteins, and mineral species like salts, metals, etc. The polarity of carbon dioxide is low, thus hindering the efficient extraction of product formed, because the extractants have a poor ability for the displacement of analyte from the matrix site. This type of media has been used for reactions catalyzed by hydrolases, particularly lipases [14–23].

The use of $SCCO_2$ as an environmentally acceptable alternative to conventional solvents for reaction chemistry, so called "Clean Technology," is being explored widely. In addition, SCFs can lead to reactions that are difficult or even impossible to achieve in conventional solvents. Their physical properties make them very attractive for biocatalytic processes. They exhibit low surface tension and viscosity compared to liquids, and high diffusivity comparable with subcritical gases; these properties favoring efficient mass transfer. On the other hand, they show liquid-like density, which promotes enhanced solubility of solutes compared to the solubility in gases. Probably the most important characteristic is that the solubility of solutes can be manipulated by changes in pressure and temperature. This makes product fractionation and purification possible directly from the reaction mixture without changing the solvent [24]. For most biocatalytic reactions, an operation temperature roughly below 333 K is required for biocatalyst stability. The choice of SCFs is thus limited to compounds with a critical temperature ($T^\wedge$ between 0 and 333 K).

3.4.1 CARBON DIOXIDE AND OTHER FLUIDS

Carbon dioxide: CO_2 is a very attractive SCF for many reasons:

   i.   very cheap and abundant in pure form (food grade) worldwide;
   ii. nonflammable and nontoxic;
   iii. environment friendly, since CO_2 can be sourced from environment pollutants such as gaseous effluents from fertilizer plants its usage can be useful in safeguarding the environment;

iv. critical temperature at 31°C, which is the optimum temperature for most biocatalysts, thus avoiding any change in the product;
v. critical pressure at 74 bar, leading to "acceptable" operation pressure, generally between 100 and 350 bar.

3.4.1.1 Biological Properties

CO_2, which is the most common SCF with low toxicity, exhibits biocidal properties against fungi, bacteria, and viruses.

Biocatalytic processes with SCFs have been limited to CO_2, except for only a few reports of bioprocesses executed with other SCFs, such as the use of polyphenol oxidase for the oxidation of p-cresol and p-chlorophenol in supercritical trifluoromethane [25] and lipase-catalyzed transesterification of methyl methacrylate in supercritical ethane, ethylene, and trifluoromethane and near-critical propane [26]. The latter study also reports for the first time on the use of an anhydrous, inorganic SCF, sulfur hexafluoride. This solvent shows the highest initial transesterification rate of all supercritical and conventional organic solvents tested. CO_2 is the most popular among the supercritical fluids because it is nontoxic, nonflammable, inexpensive, and safe for humans. $SCCO_2$ has been used as a medium for reactions catalyzed by several enzymes, for example [25, 27–39].

- alkaline phosphatase (hydrolysis of p-nitrophenyl phosphate oxidation of p-cresol and p-chlorophenol).
- polyphenol oxidase (synthesis of aspartame precursors).
- thermolysin cholesterol oxidase (oxidation of cholesterol transesterification between N-acetyl-L-phenylalanine chloroethyl ester and ethanol transesterification of triglycerides with fatty acid transesterification of methyl methacrylate with 2-ethyl hexanol).
- subtilisin (inter esterification of trilaurin and myristic acid transesterification of ethyl acetate).
- lipase (non-anol esterification of oleic acid by ethanol esterification of myristic acid and ethanol).

3.4.1.2   Stability

Enzymes generally show enhanced stability in SCFs. If stability losses are reported, they are ascribed to thermo-inactivation at the elevated temperatures used [29, 31] or to inactivation during depressurization, especially in case of enzymes with no stabilizing S-S bridges, like penicillium amidase. The degree of inactivation of monomelic enzymes with stabilizing S-S bridges, such as chymotrypsin and trypsin, during the depressurization steps is much less pronounced [40]. High moisture contents of the medium, like in organic solvents, decrease the operational stability of the enzymes. In humid CO_2, the enzyme tends to unfold more easily that further stimulates inactivation processes [41, 42].

3.5   APPLICATIONS OF SUPERCRITICAL FLUIDS

3.5.1   EXTRACTION WITH SUPERCRITICAL FLUIDS

SCFs have liquid-like densities with better mass transfer characteristics than liquid solvents due to their high diffusion and very low surface tension and are therefore used in extraction. These properties of SCF enable its easy penetration into the solid matrix with porous structure, thereby enabling the release of the solute [43, 44]. The product can be either the solid or the extract obtained, and the extracted solute can be separated from the supercritical fluid solvent by simple expansion, thus making the use of SCF more lucrative. SCF-based extraction of soluble species (solutes) from solid matrices occurs by four different mechanisms.

i.   Dissolution in suitable solvent when there are no interactions between the solute and the solid phase [45], which does not dissolves solid matrix.

ii. If there are interactions between the solid and the solute, then the extraction process is termed as desorption and the adsorption isotherm of the solute on the solid in presence of the solvent determines the equilibrium, for example, activated carbon regeneration.

iii. Extraction by swelling of the solid phase obtained by the solvent accompanied by the first two mechanisms, for example, extraction of pigments or residual solvents from polymeric matrices can also be done.

iv. When the insoluble solute reacts with the solvent, making it soluble, reactive extraction is used, such as extraction of lignin from cellulose.

Extraction by any of the abovementioned mechanism is followed by another separation process where the extracted solute is separated from the solvent [46]. Temperature, pressure, the adsorption equilibrium constant, and the solubility of the organic solvent in SCF are the thermodynamic parameters that influence the extraction in SCFs.

3.5.2   FOOD PROCESSING

CO_2 is the most common SCF in the food industry because it helps in retaining aroma, color, composition, odor, and texture of the extract and is nontoxic and has low critical temperature, thereby making it possible to be used in the extraction of thermolabile food components along with the additional advantage of the product free from residual solvent. Few applications of supercritical fluid extraction (SFE) in food are mentioned in the following subsections.

3.5.2.1   Removal of Fat from Foods

The advantages of organic solvent-free processing have led to the development of techniques, such as those involved in the removal of fat from food as in producing fat-free or fat-reduced potato chips. According to the expected taste, the amount of remaining fat in the potato chips can easily be controlled [47, 48] by using such methods.

3.5.2.2   Enrichment of Vitamin E from Natural Sources

In this method, esterification of triglycerides with methanol is done, which forms fatty acid methyl esters. These fatty acid methyl esters are easily extracted with CO_2. Solubility of free fatty acid (FFA) and tocochromanols is greater in CO_2 than in triglycerides and can be extracted as the top phase product in a separation column. This property helps in the enrichment of these components in the gaseous phase.

3.5.2.3 Removal of Alcohol from Wine and Beer, and Related Applications

Distillation has been used for this purpose for ages, but it has the disadvantage that aroma compounds are removed during the process. Use of SFE with CO_2 [47] overcomes this shortcoming.

3.5.3 ENCAPSULATION OF LIQUIDS FOR ENGINEERING SOLID PRODUCTS

A liquid product can be entrapped by methods such as

i.   adsorption onto solid particles that leads to liquid at the outside of solid particles;
ii. agglomeration in which is liquid present in the free volumes between the solid particles;
iii. impregnation in which is liquid present within the pore system of the solid particles.

Microspheres completely encapsulating the liquid and the solid material that provides a coating for the liquid inside are formed by supercritical fluid processing [48]. At optimum temperature the CO_2-liquid feed mixture (CO_2 mixed in liquid) along with the substrate is sprayed into a spray chamber. The sudden release of CO_2 from the liquid, leads to the formation of small droplets.

With this type of process, a wide variety of solid substrates can be applied to uptake liquids of different kinds [49], with advantages such as ease of handling and storage, prevention of oxidation processes, and easier dosage. Solid products can also be formed under high pressure conditions, thereby making possible the direct product formation in the SCF, which has the advantage of recycling the supercritical solvent without substantial compression [48–50]. The encapsulation or adsorption of tocopherol acetate on silica gel is one such example.

3.5.4   EXTRACTION AND CHARACTERIZATION OF FUNCTIONAL COMPOUNDS

Currently, the growing interest in the so-called functional foods having natural origin obtained by using environmentally clean techniques has raised the demand of new functional ingredients that can be used by the food industry. The complexity of the natural ingredients with biological activity is very high, thus making it necessary to develop new methodologies to extract and characterize them. Thus, techniques using compressed fluid which preserve the activity of biological compound/mixture of compounds have been developed. SFE has been used to obtain extracts with antioxidant activity from microalgae [51]. Carotenoids are a group of compounds of great importance to human health, as they can act, e.g., as potent antioxidants; however, due to their chemical characteristics, they are easily degraded by temperature or oxygen. Therefore, the use of SFE has been suggested to minimize risks of loss in activity, and thus is applied to the extraction of carotenoids from different matrices [53]. Extraction of antimicrobial compounds and food preservatives from microalgae has been accomplished using $SCCO_2$.

3.5.5   USE OF SFE IN DETERMINING FOOD SAFETY

Determination of pollutants and adulterants in food is of great importance as it is a well-known fact that addition of adulterants which are health hazards such as toxic food color. Food safety includes many different measures such as detection of frauds, adulterations, and contaminations that compromise human health. Analysis of food pollutants is commonly linked to procedures such as long extraction and clean-up, which use Soxhlet and/or saponification. Major limitations of these procedures are labor-intensive and time-consuming as they employ large volumes of toxic organic solvents.

Use of organic solvents and techniques based on compressed fluids such as SFE has wide application in analysis of pesticides and other environmental pollutants in food [54]. Several methods have been developed for the analysis of tomatoes, apples, lettuces and honey using $SCCO_2$. Similar studies have been carried out for the analysis of multiresidues of pesticides by using SFE in cereals, fish muscle, vegetable canned soups, vegetables, and infant and diet foods [54, 55] and multiple pesticides (organochlorine, organophosphorus, organonitrogen, and pyrethroid) in potatoes.

A common characteristic of these studies is the extremely high selectivity of SFE in the isolation of the low polarity pesticides; For example, pesticides and non polar pollutants can be isolated from food which could be used for its quantitative analysis.

3.6   SFE AND ANALYTICAL USES

SCCO$_2$ has been utilized in multiple methods of analysis, such as in the analyses of fat content, as a mobile phase, and in SCF chromatography. The use of SCFs are highlighted below:

3.6.1   RAPID ANALYSIS OF FAT CONTENT

SFE has been used to determine the fat content of numerous products ranging from beef to oil seeds and vegetables. The use of piezoelectric detector allows for more accurate determination of the final weight of the sample after all the fat has been extracted (total fat) facilitates the measurement of any change in the weight of the sample during the extraction process.

3.6.2   RAPID ANALYSIS OF PESTICIDES IN FOODS

Pesticide residues are a major concern among consumers throughout the world. Organic solvents such as hexane, which can extract pesticides from the sample matrix, has been used for analysis of food products and other substances such as contaminated soil and water. Once the pesticides have been extracted from the sample matrix, the samples must be "cleaned" to remove any unwanted compounds such as lipids, which may interfere with gas chromatography (GC) analysis of the sample for any pesticides present [55]. Solid phase extraction is the most common method for cleaning.

Here, SFE comes as an alternative of organic solvents for the extraction of pesticides from their sample matrix and has several advantages:

i.   the extraction can be performed in less time.
ii. the extraction utilizes less solvent volume.
iii. the extraction can be modified according to the solute of interest by altering the temperature and pressure of the extraction process.
iv. the extraction can also be modified for pesticides containing more polar groups by the addition of polar modifiers such as methanol to CO_2 [55].

3.7   NEW APPLICATIONS IN SUPERCRITICAL PARTICLE FORMATION

Apart from extraction and separation using SCFs, interest has also arisen in other areas. Here, some of the new applications of SFE are mentioned

3.7.1   NANOPARTICLE FORMATION USING SUPERCRITICAL FLUIDS

SFE have industrial applications such as preparation of micro-nano-dispersed organic systems [56]. One such example is the precipitation reaction which when done with conventional solvents needs supersaturation. Supersaturation dissolution has low dependency on temperature and it is difficult to obtain rapid heat exchange which is necessary for the process. Keeping this in mind, one process in which liquid CO_2 is used as a cooling agent becomes important where the active compound solution at −78°C is sprayed into a CO_2 stream and particle formation is induced in the spray droplets by crystallization. SF-CO_2 can act as an oxidizing agent for oxidation-sensitive compounds such as β-carotene and is thus ruled out as a precipitation medium for such compounds

3.7.2   SUPERCRITICAL DRYING

This process is similar to freeze drying to remove liquid in a controlled way, where the direct liquid-gas transition is avoided. CO_2 and freon are the suitable candidate for this purpose. It is a very commonly used technique applied for the preparation of biological samples for scanning electron microscope and in the production of aerogel. As a substance crosses the phase from being liquid to gas, it volatilizes, and thus, the volume of the liquid decreases, which causes the surface tension at the solid-liquid interface pull against any structures to which the liquid is attached.

To safeguard against breakage of delicate structures like cell walls due to dendrites in silica gel and the tiny machinery of microelectromechanical devices by surface tension, the sample can be brought from the liquid phase to the gas phase without crossing the liquid-gas boundary, instead passing through the supercritical region, where the distinction between gas and liquid ceases to apply. Nitrous oxide, though has similar physical behavior to CO_2, acts as a powerful oxidizer in its supercritical state. Supercritical water can also act as a powerful oxidizer because its critical point occurs at a temperature of 374°C and pressure of 22.064 MPa, which are very high.

3.8   STRATEGY FOR BIOCATALYSIS IN NONCONVENTIONAL MEDIA

Water is required for enzymes used in organic media to achieve good catalytic activity [57]. In esterification reactions, the initial activity of enzyme exhibits an optimum value at certain water content in the reaction media [58]. The enzymes do not dissolve in most of the commonly used organic solvents, and hence, catalysis in general takes place in dispersed media. Enzyme stability is lower in hydrophilic solvents ($-2.5 <$ log $P < 0$) such as acetone and ethers than in hydrophobic solvent ($2 <$ log $P < 4$) such as hexane, heptane, or haloalkanes still organic solvents are used. Hydrophobic solvents stabilize the biocatalyst by safeguarding the essential water layer, while hydrophilic solvents cause unfolding of the molecule by distorting this water from the enzyme surface [59–62]. Despite these obvious advantages, solvent toxicity is a problem for many applications. Successful application of biocatalysis in nonconventional media has been for ester and peptide synthesis and the resolution of chiral building blocks.

3.9   REVERSE MICELLES

Reverse micelles are very small water droplets in the range of 1–10 nm dispersed in organic media, which are obtained by the action of surfactants. They are formed in the presence of a surfactant present in a suitable concentration, forming ordered structure when a small amount of water is added to a water-immiscible hydrophobic solvent [63]. Surfactant molecules have the property to organize in such a way that polar part to the inner side facilitating to solubilize water and the nonpolar part in contact with the organic solvent. Therefore, enzyme-catalyzed reactions that require a low water content, such as those necessary to favor the synthesis reactions in organic solvents, can be achieved by microencapsulating the biocatalyst within reverse micelles. The micelles can exchange components between each other and with bulk organic solvent because of their dynamic structure. This system has a very high application in biocatalytic reactions because it can mimic the natural environment present within a cell that the enzymes encounters.

The AOT (sodium bis 2-ethyl hexyl) sulfosuccinate/isooctane system is one of the most suitable systems for enzymatic catalysis [64–66] as it provides a high interfacial area of contact with the enzyme anchoring at the aqueous side of the AOT interface, thereby enabling the use of hydrophobic substrates that are readily soluble in the organic phase [67].

The limitations of reverse micelle system include the interaction of enzyme with the micellar membrane, which can lead to changes in the micellar concentration, thus affecting the catalytic activity; in addition, dissolution of enzyme occurs in the aqueous interior pool of the micelles, which makes the activity of enzyme independent of micelle concentration. With the increase in surfactant concentration at a constant water-to-surfactant ratio, there is increase in the surface area [68]. This can be explained by taking the example of lipase enzyme. When lipase is incorporated into reverse micelle change in enzyme's secondary structure is caused due to micellar interaction. This change leads to decrease in the activity of lipase.

In ionic AOT reverse micellar system, the activity and stability of enzymes are adversely affected due to strong electrostatic interactions between AOT and enzymes [69]. However, enzymes in reverse micelles formed by some nonionic surfactants have a high activity and stability due to weak interactions between the enzyme and the nonionic surfactants [70], but the addition of some co-surfactants is still necessary for formation of the nonionic reverse micelles. A new type of surfactant, sodium bis (2-ethylhexyl polyoxyethylene) sulfosuccinate (MAOT), a chemically modified AOT, has also been prepared [71]. The chemically modified AOT (MAOT)-isooctane reversed micellar system increased substrate conversion due to decreased electrostatic and hydrophobic interactions between enzyme and MAOT molecules [72]. Such enzyme bioreactors have been used for ester synthesis [73].

3.10   IONIC LIQUIDS (ILS)

The use of ionic liquids (ILs) to improve activity, stability, and selectivity of enzymes is becoming one of the most researched areas. Ionic liquids can be defined as the organic salts that are liquid over a broad range of temperature and act as good solvents for a large variety of organic, inorganic, and polymeric compounds. The first IL reported was EtNH3 x NO3 in 1914. Presently, the most common ILs for biocatalysis are imidazolium-based ionic liquids, namely [BMIM][BF4] (1-butyl3-methylimidazolium tetrafluoroborate), [BMIM][BF6] (1-butyl-3-methylimidazolium hexafluoroborate), etc.

ILs are highly polar solvents with low melting point and are composed entirely of ions. ILs, because of their negligible vapor pressures, have been generally referred to as "green" solvents and can be used for the separation of volatile products. ILs can selectively dissolve a gas, which makes them potential solvents for gas separation [74]. CO_2 has high solubility in imidazolium-based ILs [75, 76], thus making the ionic liquids good solvents to be used for catalytic reactions such as hydrogenations, carbonylations, hydroformylations, and aerobic oxidations. Apart from having properties such as being nonvolatile (leads to reduction in pressure build-up), non-inflammable, and good thermal and chemical stability, they are immiscible with some organic solvents such as alkanes and become a potential candidate to be used in two-phase systems; thus, it is beneficial for plant design.

A huge advantage of using ILs as compared to normal organic solvents is that the physical and chemical properties of ionic liquids, including their polarity, hydrophobicity, viscosity, and solvent miscibility, can be manipulated by techniques such as altering the organic cation, inorganic anion, and attached substituents. Hence, ILs have been referred to as "designer" solvent as the flexibility to make an IL for particular reaction conditions such as alteration in enzyme selectivity, solubility of substrate or, the rate of reaction is feasible, thus making ILs an ideal solvent for engineering media for various biocatalytic reactions. The activity of enzymes in ILs were found to be better or comparable to organic solvents in some reactions [77, 78]. ILs can also be used as solvents in the extraction of a variety of substances, including metal ions, organic and biomolecules, and organosulfur from fuels and gases [79–83]. The investigations of new biphasic reactions using ILs are of special interest because it provides the possibility to adjust to the solubility properties using cation/anion combinations in various combinations [79]. Thus, biphasic reaction has been optimized with regard to product recovery and has been used to separate organic acids and carbohydrates. Moreover, recently, two groups of researchers used ILs for whole-cell in situ fermentation and showed its great potential in whole-cell biocatalytic processing due to their low toxicity to microorganisms [72, 73]. In both cases, [BMIM][PF6] was used in a two-phase system as a substrate reservoir and/or for in situ removal of the product formed, thereby increasing the productivity of the catalyst. This is useful where organic solvents used in combination with an aqueous phase either do not dissolve enough substrate or lead to increased enzyme deactivation [84]. Table 3.1 represents some recently reported examples of processes that are carried out in ILs using enzymes [85–98].

3.11 STRUCTURE AND ACTIVITY OF ENZYMES IN ORGANIC SOLVENTS

A notable change occurs in enzyme activity depending on the hydrophilic or hydrophobic nature of the organic solvent used. Measure of hydrophobicity of an organic solvent is given by partition co-efficient. It has been

TABLE 3.1 Enzyme-Catalyzed Reaction in Various Ionic Liquids (ILs)

Enzyme	ILs	Reaction	Comments
Immobilized esterases, Bacillus Stearothermophilies, Bacillus subtilis	Various ILs	Transesterification of 1-phenyl-ethanol	Higher stability of enzyme compared to organic solvents
CRL	[BMIM][BF6], [MOEMIM][BF6]	Acylation of glycosides	Higher reaction rates and selectivity than in conventional organic solvents
Immobilized CALB	[BDMIM][BF4]	Transesterification using vinyl acetate	Lipase was recycled for 10 times without losing enantioselectivity and reactivity
Immobilized CALB, α-chymotrypsin	Several ILs	Transesterification reaction	Improved the thermal stability of both the enzymes
Free Epoxide Hydrolases	Several ILs	Stereoselective hydrolysis of epoxides	Comparable reaction rate and ster selectivity than those in buffer
Immobilized PCL	[BMIM][BF6]	Resolution of racemic alcohols	Addition of triethyl amine to ILs enhanced the rate of reaction
CRL	[BMIM][BF4] [HMIM][BF4] [BMIM][BF6]	Enantioselective hydrolysis	ILs as co-solvent markedly enhanced enantioselectivity
Immobilized CALB	Several ILs	Enantioselective acylation	Increased reaction rate, decreased enantioselectivity
CRL	[BMIM][BF6], [ONIM][PF6]	Esterification of 2-substituted propanoic acids and 1-butanol	Higher enantioselectivity than in n-hexane
Immobilized PCL	[BMIM][BF4], [BMIM][BF6]	Hydrolysis and alcoholysis of 3,4,6-tri-O-actyle-D-glucal	High regioselectivity

TABLE 3.1 (Continued)

Enzyme	ILs	Reaction	Comments
CALB	Several ILs	Transesterification of ethyl butanoate with 1-butanol	Higher activity in [BMIM][BF4], [Et-3MeN][MeSO4]
Mandelate racemase, Pseudomonas putida	[MMIM][MeSO4] [BMIM][OCSO4] [BMIM][OCSO] [BMIM][O]	Kinetic resolution of mandelic acid	Reaction rate strongly influenced by aw
CAL	[BMIM][NTf2]	Kinetic resolution of rac-2-pentanol vinyl propionate	Higher activity than in hexane, greater enantioselectivity
Protease: Papain	[BMIM][BF4]	Hydrolysis of amino acid esters	Higher enantioselectivity in papain, varied enzyme activity and enantioselectivity

shown that the higher the partition co-efficient value, the more hydrophobic is the solvent. It has also been proved that enzyme activity is higher in hydrophobic solvents than in hydrophilic solvents [99–101]. This is due to the fact that the hydrophilic solvents have a tendency to eliminate (strip) tightly bound water from the enzyme molecules surface, leading to the decrease in the enzyme because the activity of the enzyme is dependent on bound water molecules. But this has no obvious effect on the backbone structure of the enzyme. Organic solvents tend to penetrate the crevices on the enzyme surface, particularly the active site when water stripping occurs, thus affecting the water activity [102, 104]. Water activity (aw) is defined as the partial vapor pressure of water in a substance divided by the standard state partial vapor pressure of water, which is an important parameter for enzyme activity in nonaqueous media; this is because enzymes require a certain level of water in their structures (bound water) to maintain their natural conformation to perform their catalytic functions [105].

While discussing enzyme flexibility, it was noted that enzyme flexibility was usually determined from molecular dynamics (MD) simulations, and it is measured by the relative calculated B-factors [106]. As is well known, non-aqueous enzymology is dependent on hydration [107]. Therefore, the water content and hydrophobicity of organic solvent have a dramatic influence on the properties of enzymes. At lower water content, the enzyme becomes rigid because it lacks flexibility, resulting in inefficient catalysis, whereas at higher water content, the enzyme starts to unfold and open up, becomes more flexible, and its activity increases. Both the conditions significantly alter its activity. Guinn et al. [108] found that the activity of horse liver alcohol dehydrogenase (HLAD) dramatically increased from 0% to 370% when water content increased from 0% to 10% in the hexane aqueous solution, but the structure of the enzyme was nearly identical to the native enzyme.

These results indicate that the water activity is essential for the activity [108]. Several studies have corroborated that alkanes (especially hexane and octane) enhance the rigidity and stability of enzymes, such as Rhizomucor miehei lipase (RML), r-chymotrypsin, subtilisin, cutinase, and horseradish peroxidase, and the level of increase is positively related to the chain length of the alkane, and thus to its hydrophobicity [109].

3.12   EFFECT OF THE FUNCTIONAL GROUPS OF ORGANIC SOLVENT

Apart from the hydrophobic or hydrophilic nature of organic solvent, the functional groups of organic solvents are also critical factors that affect enzyme activity [101]. Alkanes (such as hexane, cyclohexane, octane, and dodecane, etc.) have hydrophobic interactions with enzyme; therefore, they do not significantly change the global structure and active site of enzyme [110, 111]. For the biocatalysis in these organic solvents, the solvent molecules are located near the active site and/or close to the hydrophobic regions of the enzyme (e.g., subtilisin, cutinase, triosephosphate isomerase, etc.), resulting in the re-orientation of the side chains of some amino acids [112, 113]. Although these re-orientations of side chains do not necessarily alter the active site of the enzyme, they affect the enzyme activity by altering the substrate affinity and specificity, along with the hydration state of enzyme. For example, Pramod et al. [114] reported that although influence of octane on the secondary and tertiary structures of subtilisin was negligible or nil, the catalytic efficiency $k$cat/Km of the enzyme in octane was only 10.6% of that in aqueous solution.

This was due to the fact that the stability of subtilisin in octane was 645-fold of that in aqueous solution, owing to the absence of autolysis in octane. Similarly, Burke et al. [115] also stated that octane had little impact on the secondary structure and active site of the lytic protease; rather, it reduced the activity by altering the binding affinities of substrates to enzyme active site.

In many situations, alcohols with hydroxyl (OH) functional group are used as media for biocatalysis. The OH group results in increase in the hydrophilicity of the organic solvent. This property enhances the interactions between the solvent and enzyme. Using lipase as a model enzyme, Kamal et al. [116] demonstrated that methanol and isopropanol made lipase structure less rigid and more prone to unfolding, which increased the instability of the enzyme. The changes in enzyme structures substantially alter enzyme activities. Thus, to sum up, it can be said that hydrophobic functional groups could maintain the intact structure of enzymes so as to dramatically prolong their stability. Hydrophilic functional groups (e.g., OH, C=O, C-N, S=O, etc.) changed the enzyme structure to different extents and "stripped" the essential water from enzymes, and thereby reduced the enzyme activity.

3.13 EFFECT OF MOLECULAR STRUCTURE OF ORGANIC SOLVENT ON ENZYME ACTIVITY

The molecular structure of the organic solvent also has a great impact on the enzyme activity. This effect is due to the functional group of the solvent as well as its placement on the molecule. The organic solvent with its functional group in the terminal carbon atoms has been shown to have higher inhibitory effect on the enzyme activity than that present in the internal carbon atoms. This might be due to the possibility that organic solvents with functional groups in internal carbon atoms had higher steric effects than those in terminal carbon atoms. The higher steric effects hinder the effective interactions between these functional groups and enzyme, thus not causing much threat to the active site of the enzyme, but lowering the inhibitory effect on enzyme activity.

3.14 IONIC LIQUIDS (ILS) AS SOLVENTS

ILs, also known as molten salts, are nothing but organic salts that melt below 100°C. ILs have the advantage of having high thermal stability, negligible vapor pressure, and moderate polarity as compared to organic solvents. In addition to the abovementioned qualities, the physiochemical properties of ILs (e.g., viscosity, melting point, polarity, and hydrogen bond basicity) can be altered by making simple changes in anions or cations. Due to these advantages, ILs have presently become attractive alternatives to volatile and unstable organic solvents [117, 118].

3.15 EFFECT OF HYDROPHOBICITY OF IONIC LIQUID ON ENZYMES

The activity and stability of enzymes in the IL system can be significantly affected by the hydrophobicity of ILs [119] as discussed in the case of organic solvents. It was observed that the enzyme had higher activity and more stability in more hydrophobic ILs. Nakashima et al. [120, 121] studied the properties of PEG-modified lipase and subtilisin in three different ILs with different levels of hydrophobicity. The ILs used were 1-ethyl-3-methylimidazolium bis (trifluoromethanesulfonyl) imide, and its ether and hydroxyl analogs. They found that the activities and stabilities of the two enzymes increased with increasing hydrophobicity of the three ILs.

Similarly, Zhang et al. [122] showed that the stability of penicillin acylase was higher in more hydrophobic IL of 1-butyl-3-methylimidazolim hexafluorophosphate than in 1-butyl-3-methylimidazolium tetrafluoroborate and 1-butyl-3-methylimidazolim dicyanamide. The plausible reasoning for this phenomenon could be that increase in hydrophobicity of the ILs could lead to increase in the preservation of the essential water layer around the protein molecule, thus reducing the direct protein-ion interactions and then further enhancing the enzyme stability toward denaturing conditions [123].

Hydrophobicity of ILs could alter the selectivity of enzymes. It was claimed that enzymes showed different selectivity in the water-immiscible and water-miscible IL systems; this fact might be attributed to water activity (aw) which can alter enzyme microenvironment in IL media.

3.16   EFFECTS OF CATION AND ANION TYPES OF IONIC LIQUIDS

The types of cation and anion of ILs show great influence on enzyme activity and stability. Because of a more localized charge and stronger internal polarization of compact anion, the hydrogen bonding between enzyme and anion is much stronger than the weak van der Waals force between enzyme and cation [124]. In order to maintain the activity of IL-dissolved enzymes, a balance of mild hydrogen bond-accepting and donating property is required [125]. Therefore, anions are universally believed to exert more powerful impact on the catalytic activity and stability of enzyme than cations. This conclusion can be supported by the study by Liu et al.

However, the interactions between ILs and enzyme are complicated in practice experiments. It was speculated that the shift of enzyme activity in ILs stemmed from the secondary structure variance of enzyme, especially the alteration of α-helix and β-sheet elements [126]. ILs, in particular anions, which form strong hydrogen bonding may dissociate the hydrogen bonding that maintains the structural integrity of the α-helices and β-sheets, causing the protein to unfold wholly or partially [127]. Dabirmanesh et al. [128] demonstrated that imidazolium-based ILs could affect kinetics, structure, and stability of alcohol dehydrogenase from thermophilic *Thermoanaerobacter brockii* (TBADH).

3.17   BIOCATALYSIS IN MIXTURE SOLVENTS OF ORGANIC SOLVENT AND IONIC LIQUID

Presently, increasing attention has been focused on biocatalysis in mixture solvents of organic solvent and ILs.

3.18   SUMMARY

Enzymes in nonaqueous solutions have been largely studied and employed in the areas of food, synthesis, pharmaceuticals, and analysis. The utilization of nonaqueous solutions as reaction media could enable high enzyme activity and stability, alter enzyme selectivity, and facilitate the transformation of substrates that are unstable or poorly soluble in water. To take full advantage of the opportunities afforded by nonaqueous enzymology and develop more feasible industrial processes, this article comprehensively reviewed the structure–activity relationship of enzymes in organic solvents, ILs, sub-/supercritical fluids, and their combined mixture systems. For organic solvents and IL media, molecular interactions between the enzyme and the solvent dramatically affect the advanced structure of the enzyme and therefore its activity and stability.

Generally, solvent molecules or functional groups that interact with the enzyme through weak interactions could hold the essential bound water on the enzyme surface, stabilizing the native structure and retaining the activity of the enzyme. Unlike organic solvents, ILs also interact with the enzyme through electrostatic interaction due to the charged cation and anion. Kosmotropic anions and chaotropic cations of ILs, according to the Hofmeister series, usually act as good stabilizers of enzymes, though anions exert much greater influence on the enzyme properties owing to the strong hydrogen bonding between them. However, due to the complex interactions involved in them, it is difficult to provide a general basis for assessing the impacts of ILs on enzyme conformation and activity. For sub-/supercritical fluid medium, the main advantage for enzyme-catalyzed reactions is the tunability of solvent properties.

Therefore, the enzyme activity and product separation efficiency are dependent on temperature and pressure. It is commonly suggested that the deactivation of enzyme in sub-/supercritical fluids is caused by carbamate formation or acidification of reaction media. Interestingly, using binary mixture media of organic solvents, ILs, or sub-/supercritical fluid could effectively eliminate the demerits of the single solvent whilst preserving the merits [52]. Despite the advance of mechanisms and applications in nonaqueous enzymology, there is still much scope for improvement through research. First, more efforts should be made to understand the causes of reduced enzyme activity in nonaqueous media and how to prevent it.

The synergies of solvent engineering and protein engineering could be a potential strategy to enhance enzyme catalytic properties in nonaqueous media. Second, it is urgent to test more enzymes in nonaqueous media, especially complex enzymes. Finally, it is required to illustrate the mechanisms thoroughly and to screen more solvent-tolerant bacteria and fungal strains producing enzymes. The advances in the understanding of biocatalysis in nonaqueous systems will open a new pathway to elucidate the mechanism between structure and activity of enzymes, which will facilitate the screening of a suitable reaction medium for biotransformation.

3.19 REVIEW QUESTIONS

1. Explain what organic solvents are and list their application as non-conventional media.
2. Explain reverse micelles.
3. Explain what supercritical fluid is and list its applications.
4. What are the uses of supercritical carbon dioxide?
5. How are nanoparticles formed using supercritical fluid?
6. Explain what ionic liquids are and their applications.

KEYWORDS

- **ionic liquid**
- **non conventional media**
- **reverse micelle**
- **supercritical CO$_2$**
- **supercritical fluid**

REFERENCES

1. Marian Vermuë. *Biocatalysis in Non-conventional Media*: Kinetic and thermodynamic aspects ISBN 90-5485-462-6.
2. Kazandjian, R. Z., & Klibanov, A. M., (1985). Regioselective oxidation of phenol catalyzed by polyphenol oxidase in chloroform, *J. Am. Chem. Soc., 107,* 5448–5450.
3. Dordick, J. S., Marietta, M. A., & Klibanov, A. M., (1986). Peroxidase depolymerize lignin inorganic media, but not in water, *Proc. Natl. Acad. Sei. USA, 93,* 6255.
4. Schwartz, R. D., & McCoy, C. J., (1977). Epoxidation of 1,7-octadiene by Pseudomonas oleovorans: fermentation in the presence of cyclohexane, *J. Appl. Environ. Microbiol., 34,* 47.
5. Vermuë, M. H., & Tramper, J., (1990). Extractive biocatalysis in a liquid-impelled loop reactor. In: *Proceedings of the 5th European Congress on Biotechnology,* 243–246.
6. Zaks, A., & Klibanov, A. M., (1988). Enzymatic catalysis in nonaqueous solvents, *J. Biol. Chem. 263,* 3194–3201.
7. Zaks, A., & Klibanov, A. M., (1984). Enzymatic catalysis in organic media at 100 CC, *Science, 224,* 1249–1251.
8. Volkin, D. B., Staubli, A., et al., (1991). Enzyme thermal inactivation in anhydrous organic solvents, *Biotechnol. Bioeng., 37,* 843–853.
9. Sakurai, T., Margolin, A. L., et al., (1988). Control of enzyme enantioselectivity by the reaction medium. *J. Am. Chem. Soc., 110,* 7236–7237.
10. Klibanov, A. M., (1990). Asymmetric transformations catalyzed by enzymes in organic solvents, *Ace. Chem. Res., 23,* 114–120.
11. Chulalaksananukul, W., Condoret, J. S., & Combes, D., (1993). Geranyl acetate synthesis by lipase-catalysed transesterification in supercritical carbon dioxide, *Enzyme Microb. Technol., 15*(7), 691–698.
12. Lozano, P., Villora, G., et al., (2004). Membrane reactor with immobilized Candida Antarctica lipase B for ester synthesis in supercritical carbon dioxide, *J. Supercrit Fluids, 29*(2), 121–128.
13. Ikushima, Y., (1997). *Advances in Colloid and Interface Science, 71,* 259–280.
14. Knez, Z., & Habulin, M., (2002). Compressed gases as alternative enzymatic reaction solvents: A short review, *J. Supercrit Fluids, 23*(1), 29–42.

15. Sovova, H., & Zarevucka, M., (2003). Lipase catalysed hydrolysis of blackcurrant oil in supercritical carbon dioxide, *Chem. Eng. Sci., 58*, 2339–2350.
16. Celia, E., Cernia, E., Palocci, C., et al., (2005). Tuning pseudomonas cepacia lipase (PCL) activity in supercritical fluids, *J. Supercrit Fluids, 33*(2), 193–199.
17. Romero, M. D., Calvo, L., et al., (2005). Enzymatic synthesis of isoamyl acetate with immobilized Candida antarctica lipase in supercritical carbon dioxide, *J. Supercrit Fluids, 33*(1), 77–84.
18. Guni, T., Paolucci-Jeanjean, D., et al., (2007). Enzymatic membrane reactor involving a hybrid membrane in supercritical carbon dioxide, *J. Mem. Sci., 297*(1), 98–103.
19. Dijkstra, Z. J., Merchant, R., & Keurentjes, J. T. F., (2007). Stability and activity of enzyme aggregates of calb in supercritical CO_2, *J. Supercrit Fluids, 41*(1), 102–108.
20. Laudani, C. G., Habulin, M., Knez, Z., et al., (2007). Immobilized lipase-mediated long-chain fatty acid esterification in dense carbon dioxide: bench scale packed-bed reactor study, *J. Supercrit Fluids, 41*(1), 74–81.
21. Palocci, C., Falconi, M., et al., (2008). Lipase-catalysed regioselective acylation of tritylglycosides in supercritical carbon dioxide, *J. Supercrit Fluids, 45*(1), 88–93.
22. Verma, M. N., & Madras, G., (2008). Kinetics of synthesis of butyl butyrate by esterification and transesterification in supercritical carbon dioxide, *J. Chem. Technol. Biotechnol., 83*, 1135–1144.
23. Olivera, M. V., Rebocho, S. F., et al., (2009). Kinetic modeling of decyl acetate synthesis by immobilized lipase catalysed transesterification of vinyl acetate with decanol in supercritical carbon dioxide, *J. Supercrit Fluids, 50*(2), 138–145.
24. McHugh, M., & Krukonis, V., (1986). *Supercritical Fluid Extraction.* Butterworth Publ., Boston, USA, *13*, 1–69.
25. Hammond, D. A., Karel, M., et al., (1985). Enzymatic reactions in supercritical gases. *Appl. Biochem. Biotechnol., 11*, 393–400.
26. Kamat, S., Barrera, J., et al., (1992). Biocatalytic synthesis of acrylates in organic solvents and supercritical fluids: 1.Optimization of enzyme environment. *Biotechnol. Bioeng., 40*, 158–166.
27. Randolph, T. W., Blanch, H. W., et al., (1985). Enzymatic catalysis in a supercritical fluid. *Biotechnol. Lett., 7*, 325–328.
28. Kamihira, M., Taniguchi, M., & Kobayashi, T., (1987). Synthesis of aspartame precursors by enzymatic reaction in supercritical carbon dioxide. *Agric. Biol. Chem., 51*, 3427–3428.
29. Randolph, T. W., Blanch, H. W., & Prausnitz, J. M., (1988). Enzyme-catalyzed oxidation of cholesterol in supercritical carbon dioxide. *AIChE Journal, 34*, 1354–1360.
30. Pasta, P., Mazzola, G., Carrea, G., & Riv, S., (1989). Subtilisin-catalyzed transesterification in supercritical carbon dioxide. *Biotechnol. Lett., 2*, 643–648.
31. Nakamura, K., Chi, Y. M., Yamada, Y., & Yano, T., (1986). Lipase activity and stability in supercritical carbon dioxide. *Chem. Eng. Commun., 45*, 207–212.
32. Chi, Y. M., Nakamura, K., & Yano, T., (1988). Enzymatic inter esterification in supercritical carbon dioxide. *Agric. Biol. Chem., 52*, 1541–1550.
33. Erickson, J. C., Schyns, P., & Cooney, C. L., (1990). Effect of pressure on an enzymatic reaction in a supercritical fluid. *AIChE Journal, 36*, 299–301.
34. Kamat, S., Barrera, J., Beekman, E. J., & Russell, A. J., (1992). Biocatalytic synthesis of acrylates in organic solvents and supercritical fluids: 1.Optimization of enzyme environment. B*iotechnol. Bioeng., 40*, 158–166.

35. Miller, D. A., Blanch, H. W., & Prausnitz, J. M., (1991). Enzyme-catalyzed iner-esterification of triglycerides in supercritical carbon dioxide. *Ind. Eng. Chem. Res.,* *30,* 939–946.

36. Vermuë, M. H., Tramper, J., De Jong, J. P. J., & Oostrom, W. H. M., (1992). Enzymic transesterification in near-critical carbon dioxide: effect of pressure, Hildebrand solubility parameter and water content. *Enzyme Microb. Technol., 14,* 649–655.

37. Marty, A., Chulalaksananukul, W., Willemot, R. M., & Condoret, J. S., (1992). Kinetics of lipase catalyzed esterification in supercritical CO_2 *Biotechnol. Bioeng., 39,* 273–280.

38. Yu, Z. R., Rizvi, S. S. H., & Zollweg, J. A., (1992). Enzymatic esterification of fatty acid mixtures from milk fat and anhydrous milk fat with canola oil in supercritical carbon dioxide. *Biotechnol. Prog., 8,* 508–513.

39. Dumont, T., Barth, D., Corbier, C., Branlant, G., & Perrut, M., (1992). Enzymatic reaction kinetic: comparison in an organic solvent and in supercritical carbon dioxide. *Biotechnol. Bioeng., 39,* 329–333.

40. Kasche, V., Schlothauer, R., & Brunner, G., (1988). Enzyme denaturation in supercritical CO2: stabilizing effect of S-S bonds during the depressurization step. *Biotechnol. Lett., 10,* 569–574.

41. Weder, J. K. P., (1984). Studies on proteins and amino acids exposed to supercritical carbon dioxide extraction conditions. *Food Chem., 15,* 175–190.

42. Marty, A., Chulalaksananukul, W., Willemot, R. M., & Condoret, J. S., (1992). Kinetics of lipase catalysed esterification in supercritical CO_2 *Biotechnol. Bioeng., 39,* 273–280.

43. Van Tol, J. B. A., Stevens, R. M. M., Veldhuizen, W. J., Jongejan, J. A., & Duine, J. A., (1995). Do organic solvents affect the catalytic property of lipase? Intrinsic kinetic parameters of lipases in ester hydrolysis and formation in various organic solvents, *Biotechnol Bioeng., 47*(1), 71–81.

44. Klibanov, A. M., & Zaks, A., (1985). Enzyme-catalysed processes in organic solvents, *Proc. Natl. Acad. Sci. USA, 82,* 3192–3196.

45. Catchpole, O. J., Tallon, S. J., Grey, J. B., Fletcher, K., & Fletcher, A. J., (2008). Extraction of lipids from a specialist dairy stream. *J. Supercrit. Fluids, 45,* 314–321, DOI:10.1016/j.supflu.2008.01.004.

46. Chuang, M. H., & Brunner, G., (2005). Concentration of minor components in crude palm oil. *J. Supercrit. Fluids, 37,* 151–156, DOI:10.1016/j.supflu.09.004.

47. Bravi, E., Pperretti, G., Motanari, L., Favati, F., & Fantozzi, P., (2007). Supercritical fluid extraction for quality control in beer industry. *J. Supercrit. Fluids, 42,* 342–346.

48. Pessoa, F. L. P., & Uller, A. M. C., (2002). An economic evaluation based on an experimental study of the vitamin E concentration present in deodorizer distillate of soybean oil using supercritical CO2. *J. Supercrit. Fluids, 23,* 257265.

49. Patrick, J. G., Martin, J. W., Kevin, M. S., & Steven, M. H., (2005). Drug delivery goes supercritical. *J. Materials Today, 8,* 42–48.

50. Brunner, G., (2005). Supercritical fluids: Technology and application to food processing. *J. Food Eng., 67,* 21–33.

51. Wang, Y., Yiping, W., Yang, J., Pfeffer, R., Dave, R., & Michniak, B., Wang, Y., et al., (2006). The application of a supercritical antisolvent process for sustained drug delivery. *J. Powder Technol., 164,* 94–102.

52. Jin, Z., Han, S. Y., Zhang, L., et al., (2013). Combined utilization of lipase-displaying Pichia pastoris whole-cell biocatalysts to improve biodiesel production in co-solvent media. *Bioresour. Technol., 130*, 102–109.

53. Franceschi, E., De Cesaro, A. M., Feiten, M., Ferreira, S. R. S., Dariva, C., Kunita, M. H., Rubira, A. F., Muniz, E. C., Corazza, M. L., & Oliveira, J. V., (2008). Precipitation of β-carotene and PHBV and co-precipitation from SEDS technique using supercritical CO_2. *J. Supercritical Fluids, 47*, 256–259.

54. Rozzi, N. L., & Singh, R. K., (2002). Supercritical fluids and the food industry. *J. Comprehensive Rev. Food Sci. Food Saf., 1*, 33–44. DOI: 10.1111/j.1541–4337.2002.tb00005.x About.

55. Cortes, J. M., (2009). Pesticide residue analysis by RPLC-GC in lycopene and other carotenoids obtained from tomatoes by supercritical fluid extraction. *J. Food Chemistry, 113*, 280–284.

56. Riegr, J., & Horn, D., (2001). Organic nanoparticles in the aqueous phase-theory, experiment and use. *J. Angew. Chem. Int. Ed., 40*, 4330–4361.

57. Zaks, A., & Klibanov, A. M., (1988). Enzymatic catalysis in nonaqueous solvents, *J. Biol. Chem., 263*, 3194–3201.

58. Wehtje, E., Adlercreutz, P., & Mattiasson, B., (1990). Formation of C–C bonds by mandelonitrile lyase in organic solvents, Biotechnol. Bioeng., *36*(1), 39–46.

59. Hazarika, S., Goswami, P., Dutta, N. N., & Hazarika, A. K., (2002). Ethyl oleate synthesis by porcine pancreatic lipase in organic solvents, *Chem. Eng. J., 85*(1), 61–68.

60. Gogoi, S., Hazarika, S., Dutta, N. N., & Rao, P. G., (2006). Esterification of lauric acid with lauryl alcohol using cross-linked enzyme crystals: Solvent effect and kinetic study, *Biocatal. Biotransform., 24*(5), 343–351.

61. Sztajer, H., Lunsdorf, H., Erdmann, H., et al., (1992). *Biochim. Biophys. Acta., 1124*, 253–261.

62. Azevedo, A. M., Prazers, D. M. F., et al., (2001). Stability of free and immobilized peroxidase in aqueous-organic solvent mixtures, *J. Mol. Catal. B: Enzym, 15*(2), 147–153.

63. Tsai, S. W., & Chiang, C. L., (1991). Kinetics, mechanism and time course analysis of lipase-catalysed hydrolysis of high concentration olive oil in AOT-isooctane reverse micelles, *Biotechnol. Bioeng., 38*, 206–211.

64. Alves, J. R. S., Fonseca, L. P., Ramalho, M. T., & Cabral, J. M. S., (2003). Optimization of Penicillin acylase extraction by AOT/isooctane reversed micellar systems, *Biochem. Eng. J., 15*(1), 81–86.

65. Carvelho, C. M. L., & Cabral, J. M. S., (2000). Reverse micelles as reaction media for lipases, *Biochemie., 82*, 1063–1085.

66. Krieger, N., Tapia, M. A., Melo, E. H. M., et al., (1997). Purification of the penicillium citrinum lipase using AOT reversed micelles, *J. Chem. Technol. Biotechnol., 69*(1), 77–85.

67. Marhuenda-Egea, F. C., Piera-Velazquez, S., Cadenas, C., & Cadenas, E., (2002). Reverse micelles in organic solvents: a medium for the biotechnological use of extreme halophilic enzymes at low salt concentration, *Archaea., 1*(2), 105–111.

68. Brown, E. D., Yada, R. Y., & Marangoni, A. G., (1993). The dependence of the lipolytic activity of Rhizopus arrhizus lipase surfactant concentration in aerosol-OT/isooctane reverse micelles and its relationship to enzyme structure, *Biochim. Biophys. Acta., 1161*(1), 66–72.

69. Stamatis, H., Xenakis, A., Dimitriadis, E., & Kolisis, F. N., (1995). Catalytic behavior of pseudomonas cepacia lipase in w/o microemulsions, *Biotechnol. Bioeng.*, *45*(1), 33–41.

70. Yamada, Y., Kuboi, R., & Komasawa, I., (1993). Increased activity of chromobacterium viscosum lipase in aerosol- OT reverse micelles in the presence of non-ionic surfactants. *Biotechnol. Prog.*, *9*, 468–472.

71. Maria, P. D., Sinisterra, J. V., Montero, J. M., et al., (2006). Acyl transfer strategy for the biocatalytic characterization of Candida rugose lipases in organic solvets, *Enzyme Microb. Technol.*, *38*(2), 199–208.

72. He, Z. M., Wu, J. C., Yao, C. Y., & Yu, K. T., (2001). Lipase catalysed hydrolysis of olive oil in chemically-modified AOT/isooctane reverse micelles in a hollow fiber membrane reactor, *Biotechnol. Lett.*, *23*, 1257–1262.

73. Serralheiro, M. L. M., Prazeres, D. M. F., & Cabral, J. M. S., (1997). Continuous production and simultaneous precipitation of a di-peptide in a reversed micellar membrane reactor, *Enzym. Microb. Technol.*, *24*, 507–513.

74. Brennecke, J. F., & Maginn, E. J., (2003). Purification of gas with liquid ionic compounds, US *patent 6579, 343.*

75. Anthony, J. L., Maginn, E. J., & Brennecke, J. F., (2002). Solubilities and thermodynamic properties of gases in the ionic liquid 1-n-butyl-3-methylimidazolium hexafluorophosphate, *J. Phy. Chem. B.*, *106*, 7315–7320.

76. Cadena, C., Anthony, J. L., Shah, J. K., et al., (2004). Why is CO2 so soluble in imidazolium-based ionic liquids? *J. Am. Chem. Soc.*, *126*, 5300–5308.

77. Noel, M., Lozano, P., Vaultier, M., & Iborra, J. L., (2004). Kinetic resolution of rac-2-pentanol catalysed by Candida Antarctica lipase B in the ionic liquid, 1-butyl-3- methylimidazolium bis[(trifluoromethyl)sulfonyl]amide, *Biotechnol. Lett.*, *26*(3), 301–306.

78. Hongwei, Y., Jinchuan, W., & Bun, C. C., (2005). Kinetic resolution of ibuprofen catalysed by Candida rugose lipase in ionic liquids, *Chirality, 17*(1), 16–21.

79. Zhao, H., Shuqian, S., & Ma, P., (2005). Use of ionic liquids as green solvents for extractions, *J. Chem. Technol. Biotechnol.*, *80*, 1089–1096.

80. Chun, S., Dzyuba, S. V., & Bartsch, R. A., (2001). Influence of structural variation in room-temperature ionic liquids on the selectivity and efficiency of comparative alkali metal salt extraction by a crown ether, *Analytical Chemistry, 73*, 3737–3741.

81. Dai, S., Ju, Y. H., & Barnes, C. E., (1999). Solvent extraction of strontium nitrate by a crown ether using room temperature ionic liquids, *Journal of the Chemical Society, Dalton Transactions, 8*, 1201–1202.

82. Vidal, S., Neiva, C. M. J., Marques, M. M., Ismael, M. R., & Angelino, M. T., (2004). Studies on the use of ionic liquids as potential extractants of phenolic compounds and metal ions, *Separation Science and Technology, 39*, 2155–2169.

83. Matsumoto, M., Mochiduki, K., Fukunishi, K., & Kondo, K., (2004). Extraction of organic acids using imidazolium-based ionic liquids and their toxicity to Lactobacillus rhamnosus, *Sep. Purif. Technol.*, *40*(1), 97–101.

84. Wasserscheid, P., & Welton, T., (2002). *Ionic Liquids in Synthesis.* Wiley-VCH, Weinheim.

85. Persson, M., & Bornscheuer, U. T., Increased stability of an esterase from Bacillus stearothermophilus in ionic liquids as compared to organic solvents, *J. Mol. Catal. B: Enzym.*, *22*(1), 21–27.

86. Kim, M. J., Choi, M. Y., Lee, J. K., & Ahn, Y., (2003). Enzymatic selective acylation of glycosides in ionic liquids: significantly enhanced reactivity and regioselectivity, *J. Mol. Catal. B: Enzym.*, *26*(2), 115–118.

87. Itoh, T., Nishimura, Y., Ouchi, N., & Hayase, S., (2003). 1-butyl-2,3- dimethylimidazolium tetrafluoroborate: the most desirable ionic liquid solvent for recycling use of enzyme in lipase –catalysed transesterification using vinyl acetate as acyl donor, *J. Mol. Catal. B: Enzym.*, *26*(1), 41–45.

88. Lozano, P., Diego, T. D., Carrie, D., Vaultier, M., & Iborra, J. L., (2003). Enzymatic ester synthesis in ionic liquids, *J. Mol. Catal. B: Enzym.*, *21*(1), 9–13.

89. Kaar, J. L., Jesionowski, A. M., Berberich, J. A., Moulton, R., & Russell, A. J., (2003). Impact of ionic liquid physical properties on lipase activity and stability, *J. Am. Chem. Soc.*, *125*, 4125–4131.

90. Rasalkar, M., Potdar, M. K., & Salunke, M. M., (2004). Pseudomonas cepacia lipase-catalysed resolution of racemic alcohols in ionic liquid using succinic anhydride: role of triethylamine in enhancement of catalytic activity, *J. Mol. Catal. B: Enzym.*, *27*, 267–270.

91. Mohile, S. S., Potdar, M. K., Harjani, J. R., Nara, S. J., & Salunke, M. M., (2004). Ionic liquids: efficient additives for Candida rugosa lipase-catalysed enantioselective hydrolysis of butyl 2-(4-chlorophenoxy)propionate, *J. Mol. Catal. B: Enzym.*, *30*(2), 185–188.

92. Irimescu, R., & Kato, K., (2004). Lipase catalysed enantioselective reaction of amines with carboxylic acids under reduced pressure in non-solvent system and in ionic liquids, *Tetrahedron Lett.*, *45*, 523–525.

93. Ulbert, O., Frater, T., Belafi-Bako, K., & Gobicza, L., (2004). Enhanced enantioselectivity of Candida rugosa lipase in ionic liquids as compared to organic solvents, *J. Mol. Catal. B: Enzym.*, *31*(1), 39–45.

94. Nara, S. J., Mohile, S. S., Harjani, J. R., Naik, P. U., & Salunke, M. M., (2004). Influence of ionic liquids on the rates and regioselectivity of lipase-mediated biotransformations on 3,4,6-tri-O-acetyl glucal, *J. Mol. Catal. B: Enzym.*, *28*(1), 39–43.

95. Lau, R. M., Sorgedrager, M. J., Carrea, G., Rantwijk, F. V., Secundo, F., & Sheldon, R. A., (2004). Dissolution of Candida antarctica lipase B in ionic liquids: effects on structure and activity, *Green Chem.*, *6*, 483–487.

96. Kaftzik, N., Kroutil, W., Faber, K., & Kragl, U., (2004). Mandelate racemase activity in ionic liquids: scopes and limitations, *J. Mol. Catal. A: Chem.*, *214*(2), 107–112.

97. Noel, M., Lozano, P., Vaultier, M., & Iborra, J. L., (2004). Kinetic resolution of rac-2-pentanol catalysed by Candida antarctica lipase B in the ionic liquid, 1-butyl-3-methylimidazolium bis [(trifluoromethyl)sulfonyl]amide, *Biotechnol. Lett.*, *26*, 301–306.

98. Liu, Y. Y., Lou, W. Y., Zong, M. H., et al., (2005). Increased enantioselectivity in the enzymatic hydrolysis of amino acid esters in the ionic liquid 1- butyl-3-methylimidazolium tetrafluoroborate, *Biocatal. Biotransform.*, *23*(1), 89–95.

99. Liu, Y., Zhang, X., Tan, H., et al., (2010). Effect of pretreatment by different organic solvents on esterification activity and conformation of immobilized pseudomonas cepacia lipase. *Process Biochem.*, *45*, 1176–1180.

100. Su, E., & Wei, D., (2008). Improvement in lipase-catalyzed methanolysis of triacylglycerols for biodiesel production using a solvent engineering method. *J. Mol. Catal. B.*, *55*, 118–125.

101. Liu, Y., Tan, H., Zhang, X., Yan, Y., & Hameed, B., (2010). Effect of monohydric alcohols on enzymatic transesterification for biodiesel production. *Chem. Eng. J., 157*, 223–229.

102. Trodler, P., & Pleiss, J., (2008). Modeling structure and flexibility of Candida Antarctica lipase B in organic solvents. *BMC Struct. Biol., 8*(9).

103. Secundo, F., Fiala, S., Fraaije, M. W., et al., (2011). Effects of water miscible organic solvents on the activity and conformation of the Baeyer-Villiger monooxygenases from Thermo bifida fusca and Acinetobacter calcoaceticus: A comparative study. *Biotechnol. Bioeng., 108*, 491–499.

104. Rezaei, K., Jenab, E., & Temelli, F., (2007). Effects of water on enzyme performance with an emphasis on the reactions in supercritical fluids. *Crit. Rev. Biotechnol., 27*, 183–195.

105. Fasoli, E., Ferrer, A., & Barletta, G. L., (2009). Hydrogen/deuterium exchange study of subtilisin Carlsberg during prolonged exposure to organic solvents. *Biotechnol. Bioeng., 102*, 1025–1032.

106. Pérez-Castillo, Y., Froeyen, M., Cabrera-Pérez, M. A., & Nowé, A., (2011). Molecular dynamics and docking simulations as a proof of high flexibility in E. coli FabH and its relevance for accurate inhibitor modeling. *J. Comput. Aided. Mol. Des., 25*, 371–393.

107. Lousa, D., Baptista, A. M., & Soares, C. M., (2013). A molecular perspective on nonaqueous biocatalysis: Contributions from simulation studies. *Phys. Chem. Chem. Phys., 15*, 13723–13736.

108. Mattos, C., Bellamacina, C. R., Peisach, E., et al., (2006). Multiple solvent crystal structures: Probing binding sites, plasticity and hydration. *J. Mol. Biol., 357*, 1471–1482.

109. Yang, L., Dordick, J. S., & Garde, S., (2004). Hydration of enzyme in nonaqueous media is consistent with solvent dependence of its activity. *Biophys. J., 87*, 812–821.

110. Lousa, D., Baptista, A. M., & Soares, C. M., (2013). A molecular perspective on nonaqueous biocatalysis: Contributions from simulation studies. *Phys. Chem. Chem. Phys., 15*, 13723–13736.

111. Yennawar, N. H., Yennawar, H. P., & Farber, G. K., (1994). X-ray crystal structure of gamma-chymotrypsin in hexane. *Biochemistry, 33*, 7326–7336.

112. Wangikar, P. P., Michels, P. C., Clark, D. S., & Dordick, J. S., (1997). Structure and function of subtilisin BPN' solubilized in organic solvents. *J. Am. Chem. Soc., 119*, 70–76.

113. Burke, P. A., Smith, S. O., Bachovchin, W. W., & Klibanov, A. M., (1989). Demonstration of structural integrity of an enzyme in organic solvents by solid-state NMR. *J. Am. Chem. Soc., 111*, 8290–8291.

114. Guinn, R. M., Skerker, P. S., Kavanaugh, P., & Clark, D. S., (1991). Activity and flexibility of alcohol dehydrogenase in organic solvents. *Biotechnol. Bioeng., 37*, 303–308.

115. Choi, Y. S., & Yoo, Y. J., (2012). A hydrophilic and hydrophobic organic solvent mixture enhances enzyme stability in organic media. *Biotechnol. Lett., 34*, 1131–1135.

116. Kamal, M. Z., Yedavalli, P., Deshmukh, M. V., & Rao, N. M., (2013). Lipase in aqueous-polar organic solvents: Activity, structure, and stability. *Protein Sci., 22*, 904–915.

117. Mora-Pale, M., Meli, L., Doherty, T. V., Linhardt, R. J., & Dordick, J. S., (2011). Room temperature ionic liquids as emerging solvents for the pretreatment of lignocellulosic biomass. *Biotechnol. Bioeng., 108*, 1229–1245.

118. Gorke, J., Srienc, F., & Kazlauskas, R., (2010). Toward advanced ionic liquids. Polar, enzyme-friendly solvents for biocatalysis. *Biotechnol. Bioprocess Eng., 15*, 40–53.

119. Weingärtner, H., Cabrele, C., & Herrmann, C., (2012). How ionic liquids can help to stabilize native proteins. *Phys. Chem. Chem. Phys., 14*, 415–426.

120. Nakashima, K., Maruyama, T., Kamiya, N., & Goto, M., (2006). Homogeneous enzymatic reactions in ionic liquids with poly (ethylene glycol)-modified subtilisin. *Org. Biomol. Chem., 4*, 3462–3467.

121. Nakashima, K., Okada, J., Maruyama, T., Kamiya, N., & Goto, M., (2006). Activation of lipase in ionic liquids by modification with comb-shaped poly (ethylene glycol). *Sci. Technol. Adv. Mater., 7*, 692–698.

122. Zhang, W. G., Wei, D. Z., Yang, X. P., & Song, Q. X., (2006). Penicillin acylase catalysis in the presence of ionic liquids, *Bioprocess Biosyst. Eng., 29*, 379–383.

123. Lozano, P., De Diego, T., Guegan, J. P., Vaultier, M., & Iborra, J. L., (2001). Stabilization of chymotrypsin by ionic liquids in transesterification reactions. *Biotechnol. Bioeng., 75*, 563–569.

124. Li, N., Du, W., Huang, Z., Zhao, W., & Wang, S., (2013). Effect of imidazolium ionic liquids on the hydrolytic activity of lipase. *Chin. J., Catal., 34*, 769–780.

125. Lau, R. M., Sorgedrager, M. J., Carrea, G., et al., (2004). Dissolution of Candida Antarctica lipase B in ionic liquids: Effects on structure and activity. *Green Chem., 6*, 483–487.

126. Galonde, N., Nott, K., Richard, G., et al., (2013). Study of the influence of pure ionic liquids on the lipase-catalyzed (trans) esterification of mannose based on their anion and cation nature. *Curr. Org. Chem., 17*, 763–770.

127. Van Rantwijk, F., Secundo, F., & Sheldon, R. A., (2006). Structure and activity of Candida Antarctica lipase B in ionic liquids. *Green Chem., 8*, 282–286.

128. Dabirmanesh, B., Khajeh, K., Ranjbar, B., Ghazi, F., & Heydari, A., (2012). Inhibition mediated stabilization effect of imidazolium based ionic liquids on alcohol dehydrogenase. *J. Mol. Liquids 170*, 66–71.

CHAPTER 4

METABOLIC ENGINEERING

ANJALI PRIYADARSHINI and PRERNA PANDEY

CONTENTS

4.1   INTRODUCTION TO METABOLIC ENGINEERING

Metabolic engineering is a new field with applications in the production of chemicals, fuels, materials, pharmaceuticals, and medicine at the
genetic level. The field's novelty is in the synthesis of molecular biology
techniques and the tools of mathematical analysis, which allow rational

selection of targets for genetic modification through measurements and control of metabolic fluxes.

Primary aim of this exercise is to identify specific genetics or environmental manipulations that result in improvements in yield and productivities of biotechnological processes. Metabolic engineering is a very advance technique that is widely being used for optimization and introduction of new cellular processes by genetic engineering. This is achieved firstly by proposing the metabolic models.

Subsequently, modification is done followed by analysis of recombinant strains. The results thus obtained are used to identify the next target for genetic manipulation if needed. Thus, it leads to generation of optimal strains for successful operation. Penicillin was the first antibiotic to be discovered as produced by microorganism. Although a wide variety of microorganisms synthesize antibiotics, the majority of clinically useful antibiotics are produced by the eubacteria Actinomycetes, in particular *Streptomyces*, and the filamentous fungi. β-lactams are among the important antibiotics acting against a wide range of pathogens. As already known, most antibiotics are secondary metabolites that are synthesized from primary metabolite precursors, and they are not associated with the viability of the producer. Thus, this pathway can be used for metabolic engineering to alter the organism. This could lead to the synthesis of new or known antibiotic(s) in an organism that is already a producer or in the one that was not producing at all.

Besides, by metabolic engineering, an organism can also be altered to synthesize an intermediate that previously had to be supplied exogenously for the generation of desired product, thus easing out the manufacturing process (Figure 4.1). Metabolic engineering can also be used to eliminate the biosynthesis of undesirable by-products. It is also possible to change

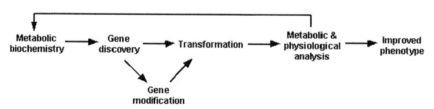

FIGURE 4.1   The basic layout of modification of organism for metabolic engineering.

flux distributions of compounds involved in antibiotic biosynthesis to increase the yields or productivity. Though a lot of interest has been generated in this field and a lot is being done, the results are few owing to various reasons:

- The complicated biosynthetic pathways leading to the final product and the relatively slow development of genetic tools.
- Most of the biochemical pathways involved in antibiotic biosynthesis are not completely characterized, the enzymes involved are not isolated, and the loci of genes encoding these enzymes are not known.
- Furthermore, most secondary metabolites are produced by microorganisms that undergo differentiation in their growth cycle. The regulation of gene expression for the biosynthesis of secondary metabolites and its relation to differentiation is little known.
- The interface between the primary and secondary metabolism is also poorly understood.

The main goals of metabolic engineering are:

- Improvement of yield, productivity, and overall cellular physiology.
- Extension of the substrate range.
- Deletion or reduction of by-product formation.
- Introduction of pathways forming new products.

In traditional chemical processes, a low-value starting material is converted into a high-value product through a series of unit operations. Initial operations may concentrate or refine the starting material by separating it from contaminants. The processed starting material is reacted with additional substrates in the presence of a catalyst, and the product of interest is separated from unreacted substrates and byproducts. Advances in catalysis and process optimization maximize single-pass conversion and profitability.

Microbial cell factories (MCFs) have emerged as a revolutionary platform for combining traditional unit operations and complex multistep catalysis into a single self-replicating microbe [1]. Reactors filled with billions of microbes can now replace much of the traditional chemical factory. Each cell can selectively uptake a low value substrate and use its vast metabolic network (and compartmentalization if necessary) to produce desired products (Figure 4.2). CRISPR-Cas9 is a simple and efficient tool for targeted and marker-free genome engineering. The later part of this chapter deals with the recent advances in

- How the microbes are selected and genetically modified (engineered) to serve as an MCF.
- How new catalytic properties are added to the metabolic network to obtain desired product.
- How the cell is engineered to use new metabolic pathways to maximize yield of a desired product.

Before elaborating on MCFs, it becomes pertinent to discuss the role of plant cell in metabolic engineering, because besides microbial cell, plant cell too provides a platform for metabolic engineering, which can be harnessed to modify and upscale the desired product. Plants provide a very large pool of structurally diverse chemicals. The biosynthesis of such compounds occurs in a response to external or environmental stimulus, which plays a very important role in shaping the interdependence and diversity of plant ecosystems. These chemicals have a widespread effect on plants.

Apart from having an effect on plants, many chemicals produced by plants are known to promote human health. Multiple plant metabolites have been isolated for use in the pharmaceutical industry. Despite the importance of plant metabolites, not much is known about the complex

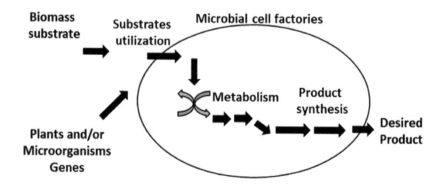

FIGURE 4.2 Development of microbial cell factory to upscale and optimize the production of desired substance.

biosynthetic processes of all, thus indicating that the immense potential in the field of plant metabolism needs to be explored. The growth in this field has been facilitated by the recent advances in next-generation sequencing technologies as well as with the continuous development of new algorithms for bioinformatic analysis of these sequence data. By extension, these discoveries have allowed advancements in the engineering of plant metabolism for economic gain.

It is of great importance to elucidate and engineer the plant metabolic pathways that form complex metabolites fusing simple building blocks, for example, starch is made up of simple glucose molecules. An understanding of these pathways will allow us to fully harness the wealth of compounds and biocatalysts that plants provide. A wide range of problems can be addressed by plant metabolic engineering.

4.2   METABOLIC ENGINEERING IN PLANTS

Plants contain numerous metabolic pathways that are responsible for the biosynthesis of complex metabolites. They act as an inexhaustible source of structurally diverse chemicals. Biosynthesis of these compounds is a response to external or environmental stimuli. These chemicals impact how effectively plants can be used as food and energy sources. Moreover, many chemicals that are produced by plants are used for aiding human health as well as for use in the pharmaceutical industry.

4.2.1   INCREASING FOOD SECURITY BY INTRODUCTION OF NEW TRAITS IN PLANT METABOLISM

With the ever-increasing population of the Earth, food security is a tremendously important issue: we cannot `increase the cultivable land; thus, the only way left to enhance the food security is by obtaining the maximum nutritional value from the crops. As already known, plant produces a plethora of metabolites having important nutritional and health benefits. Thus, by modifying and upregulating the pathways for these metabolites, crops can be made more nutritionally sounder. For example, phenylpropanoids are plant metabolites that have the property to act as antioxidant agents, and thus have very essential health promoting properties [2]. Thus, metabolic engineering of phenylpropanoid, which play a very important role in human diet, would greatly impact world food security [2].

While engineering the increase in the levels of these compounds, it is important to increase the level in edible parts of crop plants to impact human nutrition. Tomato plant has been put to great use for such engineering purposes. As phenylpropanoid has a great nutritional value, its production has been enhanced/upregulated by introducing fruit-specific expression of the *Arabidopsis thaliana* transcription factor AtMYB12 [3]. The transcription factor AtMYB12 increases phenylpropanoid levels by transcriptionally activating the biosynthetic genes of the pathways. Besides this, the transcription factor also has a role in direct carbon flux toward aromatic amino acid biosynthesis, which in turn increases the supply of substrate for phenylpropanoid metabolism. Betalains are widely used as natural food colorants and dietary supplements [4] and L-DOPA, a betalain pathway intermediate, is widely used for the treatment of Parkinson's disease [5]. Most notably, the first committed step in the pathway, 3-hydroxylation of tyrosine to form L-3,4-dihydroxyphenylalanine (L-DOPA), is not characterized.

4.2.2 ENGINEERED PLANTS AS ENVIRONMENTAL CLEANSERS

Heavy metals pose to be a great environmental threat, which is also referred to as metal toxicity. Thus, metabolic engineering process comes as a rescue [6]. For example, cadmium if present in the environment binds to the thiol groups of proteins and coenzymes and leads to displacement of endogenous metal cofactors from native binding partners, thus affecting their function. Certain peptides that have the function to protect the plant from metal toxicity are known as phytochelatins, which act by binding tightly to such metals.

4.2.3 ENGINEERING PLANT PATHWAYS TO CREATE BETTER BIOFUELS

Need of time is to make a shift from fossil-based energy sources to biofuels and the transition to a bio-based sustainable economy. Development of efficient and cost-effective biofuels remains a very big challenge.

Bioethanol is one of the prominent example of the major biofuel currently in use. Production of ethanol is done by using easily available accessible sugars of sugarcane and corn as substrate. The major drawback of this is the issue of food security, as these plants also double as very important food source; thus, it becomes pertinent to look for alternative methods for biofuel production [7]. Efforts have been made to produce the next generation of biofuels from the biomass which do not have other nutritional or economic advantage. Lignocellulose, which originates from residual biomass of crops such as wheat, corn, and sugarcane, has emerged as a promising source. Otherwise, the biomass from crops such as poplar and switchgrass that can be grown on marginal land are also possibilities for fuel production

Plant cell walls contain a large amount of lignin that affects the access of enzymes to polysaccharides of biomass which is required. To take care of this biomass plant cells are first subjected to acidic or alkaline hydrolysis to break the bond between lignin and hemicellulose, followed by enzymatic degradation, so to be used as biomass. Therefore, there has been a substantial effort in metabolic engineering to reduce lignin content in plants, because it acts as the major limiting factor of conversion of biomass to fermentable sugars. One recent study exploited a key enzyme in lignin biosynthesis, cinammoyl-CoA reductase (CCR), which catalyzes the conversion of hydroxycinnamoyl-CoA esters to the corresponding aldehydes [8]. Field trials on poplar plants have shown that biomass from transgenic plants with downregulation of CCR is more easily processed to produce bioethanol.

Another source of biofuel can be storage lipids in plants, triacylglycerols (TAGs), which are one of the most abundant and energy-rich forms of reduced carbon in nature, which can be readily converted to biofuels. A second strategy to improve access to biofuels can be to increase the content of TAGs in plant vegetative tissues. A variety of genes that enhance TAG accumulation levels have been identified: the transcription factor WRINKLED1, the TAG biosynthetic gene diacylglycerol acyltransferase1-2 (*DGAT1-2*), and a gene encoding a structural protein oleosin1 (*OLE1*) that impacts oil body formation. Moreover, it has been shown that silencing an enzyme involved in starch biosynthesis, ADP-glucose pyrophosphorylase (AGPase), diverts carbon away from starch and into TAG biosynthesis, and silencing of the peroxisomal ABC transporter1 (PXA1) prevents fatty acids from being oxidized in the mitochondria. All this information can be readily used for enhancing the production of biofuel.

4.2.4   THE NEXT GENERATION OF ENGINEERING PLANT METABOLIC PATHWAYS

While metabolic engineering of plant pathways has made substantial leaps in the last several years, new approaches to manipulate plant pathways are continually emerging. Perhaps most notably, the CRISPR-Cas9 genome engineering system has become an important new genome-editing tool for plant biologists due to this system's efficiency and specificity [9]. While CRISPR-Cas9 studies in plants have been largely confined to proof-of-concept studies, the approach has been implemented in a number of eco-nomically important crop plants [10].

4.3   MICROBIAL CELL FACTORIES (MCFS)

4.3.1   CONVENTIONAL METHODS OF STRAIN IMPROVEMENT

Since the discovery of antibiotics, their production has been enhanced to multiple times majorly owing to improvements done in the producing strain apart from medium development and process engineering. It is also through strain improvement that the producing strains have been adapted to economically desirable substrates and/ or to eliminate undesirable by-products. Strain improvement program is a two-step process:

    i. Generation of genotype variants in the population using:
- physically induced mutation
- chemically induced mutations
- by recombination among strains.
   ii. Selection or screening of those with improved phenotype properties.

Enhancement of antibiotic production in an improved strain can be through various factors which are majorly:

- An increased flux of a precursor primary metabolite may lead to an increased productivity.
- Some antibiotics are toxic even to the producing microorganisms; therefore, an increased resistance of the producer to the antibiotics can lead to an enhanced productivity.
- Another possible mechanism for increased production is to enhance gene expression and the resulting concentrations of enzymes involved in antibiotic biosynthesis.
- Besides random screening for mutants with the desired phenotype properties, selection strategies can be designed to facilitate the isolation of mutants having one of the above mechanisms.

For example, selection of mutants resistant to analogs involved in the precursor synthesis may enhance the probability of isolating a mutant that is deregulated in the biosynthesis of the precursor and overproduces it.

Another method used is the selection of auxotrophs of the precursor followed by isolating a revertant of the prototrophic microorganism. In a strain used in commercial production, often both the precursor fluxes and the antibiotic biosynthetic machinery have been greatly enhanced.

All the mechanisms contributing to the increase in antibiotic production as described above can be exploited by metabolic engineering. However, because metabolic engineering of strains requires to manipulate the genes involved in the antibiotic biosynthesis, detailed knowledge of the biochemical pathways, the genes encoding for the enzymes, and the regulation of the expression of these genes becomes beneficial. In addition to such biochemical knowledge and the availability of genetic tools, the employment of various analytical tools for determining the bottlenecks of the biosynthesis is also important. Thus, rational and computational tools have become handy for such processes.

4.3.2   TOOLS OF METABOLIC ENGINEERING

4.3.2.1   Availability of Mutants

With availability of such mutants using labeled precursors, which have been blocked for accumulation of reaction intermediates in the culture medium, the identity of the intermediates, and the knowledge on chemical conversions of these compounds, it is possible to elucidate the various biochemical pathway. In the last two decades, the availability of the various genetic tools has greatly facilitated the elucidation of the biochemical pathway of secondary metabolite synthesis.

4.3.2.2   Clustered Genome

The fact that many genes involved in the biosynthesis of an antibiotic, or a group of antibiotics, were often clustered in the genome of microorganism has eased the elucidation of pathways significantly and thus facilitated metabolic engineering.

4.3.2.3   Reverse Genetics

The biosynthetic genes have often been found to be clustered with the resistance genes that protect the microorganism from the effect of antibiotic. Techniques used in identification of the genes involved include screening of cosmids via degenerate synthetic oligonucleotide probes (multiple nucleotide sequences each of which codes for the same peptide sequence) based on partial amino acid sequence of biosynthetic enzymes. This approach is sometimes termed as "reverse genetics."

4.3.2.4   Use of Heterologous Probes

It is possible to employ heterologous probes that include conserved sequence of genes in similar pathways in another organism. This allows one to locate the genes locus of pathways that are hypothesized to include evolutionarily related enzymes.

4.3.2.5   Complementation

Complementation of blocked mutants with a gene library from the original wild type can also be used to identify the responsible genes.

Because most antibiotic biosynthetic genes have been found to be clustered together with resistance-encoding genes, another successful technique has relied on cloning of the resistance gene fragment by phenotype selection in a heterologous host. This is then followed by screening of library clones to identify biosynthetic genes. Screening for heterologous expression of the entire antibiotic pathway via cosmids in another strain with a null background has also been used (Figure 4.3).

Another approach is use of the mutational cloning technique where random disruptions are done and clones are selected for disruption in antibiotic biosynthesis. Differential screening of the transcripts produced during the exponential versus the stationary phase has been utilized to isolate eukaryotic penicillin biosynthetic genes active in the stationary phase of growth. By using all these techniques, the number of pathways elucidated have increased [11–14]; thus now, it has become easier to use the available information and techniques to decipher unknown pathways rapidly.

4.4 REGULATION OF BIOSYNTHETIC PATHWAY

Regulation of secondary metabolism is a very complex process occurring at both genetic and biochemical level. Most of the microorganisms producing secondary metabolites undergo differentiation during their life cycle. Thus, the onset of secondary metabolism is often associated with the differentiation events.

In *Streptomyces,* a hierarchy in the regulation of antibiotic biosynthesis and differentiation has been observed.

- At the global level, the production of secondary metabolite may be coregulated with other differentiation phenomena such as sporulation; a global regulatory gene which turns on sporulation possibly

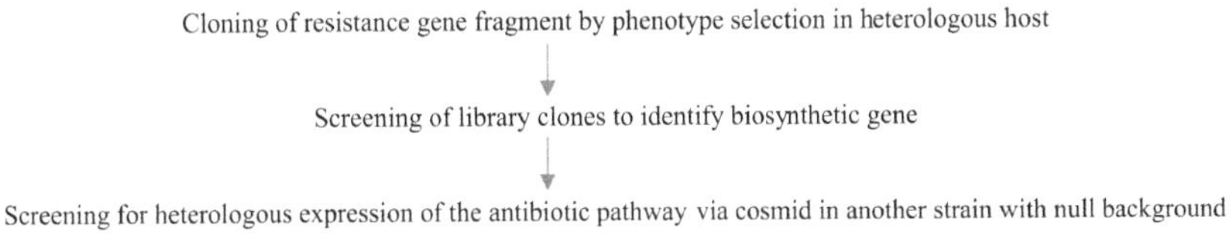

Cloning of resistance gene fragment by phenotype selection in heterologous host

↓

Screening of library clones to identify biosynthetic gene

↓

Screening for heterologous expression of the antibiotic pathway via cosmid in another strain with null background

FIGURE 4.3 Flowchart showing cloning of the resistance gene fragment by phenotype cloning of resistance gene fragment by phenotype selection in a heterologous host.

also turns on the production of secondary metabolites [15].

- In *Streptomyces,* which produce multiple secondary metabolites, pleiotropic regulators control some or all of these biosynthetic pathways.
- Lowermost in the hierarchy of regulation are the pathway specific activators, which control the production of only a single antibiotic, or a number of related antibiotics that share segments of the same biosynthetic pathway.
- In addition to pathway-specific activators, antibiotic biosynthesis in some cases is also regulated by pathway-specific repressors. Although most regulatory molecules that control production are intracellular, there are a few extracellular effector molecules with hormone-like properties. Notable examples are A-factor in *S. griseus* [16] and Virginia butanolides in *Streptomyces virginiae* [17]. A-factor and Virginia butanolides are autoregulators whose synthesis turns on their own production and triggers both sporulation and antibiotic production. They are excreted extracellularly to induce antibiotic production and sporulation in neighboring cells.

4.4.1 AT THE GENETIC LEVEL

- Regulation involves the induction or repression of the expression of the enzymes in the biosynthetic pathways. Available evidence suggests that the activity of the enzymes is coordinately regulated by the pathway-specific regulators [18].
- The regulatory circuit controls not only the level of enzyme synthesized but also the temporal profiles of these enzymes, i.e., when they are synthesized and when they are turned over to shut down biosynthesis.
- Although evidence is scarce, it is conceivable that regulation occurs also at the spatial level, i.e., only certain portions of the organisms are actively producing secondary metabolite. The temporal regulation of secondary metabolism is well known; indeed, it is such a profile where metabolite synthesis lags behind the initial rapid expansion of biomass that gives the term "secondary metabolism."

The regulation at the biochemical level involves the control of enzyme activity including feedback inhibition, activation, and inactivation. Another thing to be determined is the rate limiting step for the biochemical reaction. The mechanism through which this is achieved has been elaborated in Figure 4.4. The β-lactam antibiotic biosynthesis is a very appropriate example to show the effect of branched biosynthetic pathways on antibiotic production. The three precursors needed for β-lactam antibiotics from primary metabolism are L-Lysine, L-Cysteine, and L-Valine, which are also needed for cellular protein synthesis as a primary requirement of the cell. The competing reactions for these amino acids therefore involve the binding of these amino acids to tRNA aided by particular aminoacyl-tRNA synthetases. The Km for the binding of the amino acid to the tRNA is usually in the range of micromolar, whereas the Km values for these amino acids for β-lactam biosynthetic enzymes

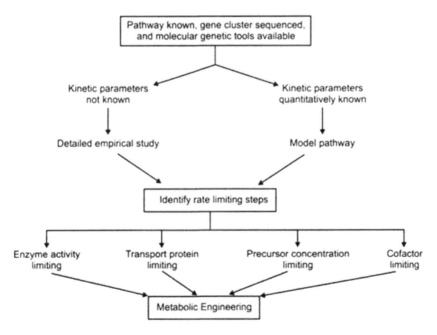

FIGURE 4.4   A well-characterized biosynthetic pathway that aids and allows prediction of the rate limiting steps.

(L-Lysine aminotransferase (LAT) and ACV synthetase) are in millimolars. Because the Km for these amino acids for secondary metabolism are at least an order of magnitude higher than those for protein synthesis, diversion of precursors preferentially toward secondary metabolism at the expense of primary metabolism is not favored. Moreover, microorganisms also control the flux of precursors and cofactors into secondary metabolism by a variety of means to ensure that these precursors and cofactors are not diverted when they are needed for cell growth. This type of control differs from conventional feedback inhibition, repression, and induction in that it is affected by the balance of fluxes in a metabolic network rather than involving only a single biosynthetic pathway. This makes the process of metabolic engineering difficult. These are the hurdles that need to be overcome. Using a stoichiometric model that accounts for the major reactions, the fluxes through various pathways can be solved for a given pseudo steady state. This can then give information on the expected yields, which could be the yardstick for directing the metabolic engineering efforts.

4.5 EXAMPLES OF METABOLIC ENGINEERING OF ANTIBIOTIC BIOSYNTHESIS

4.5.1 INCREASING THE METABOLIC FLUX

There are many ways in which secondary metabolite production by metabolic engineering could be enhanced. The production of these metabolite is not a single step process, but it is multistep that involves the production of certain precursor molecules prior to the production of the secondary metabolite. Thus, if the precursor to the metabolite is provided in the culture media, the organism does not need to spend energy on its production and straight away can form the product from the precursor provided at an increased rate and concentration. As in the case of antibiotic production that requires many cofactors, if provided can lead to amplification of one or more enzymes involved in the biochemical pathways and improving the kinetic characteristics of the enzyme (Figures 4.4 and 4.5).

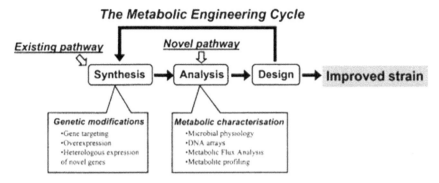

FIGURE 4.5 Various methods employed for strain improvement for modification in existing pathways to design a novel pathway through metabolic engineering.

4.5.1.1 Enhancement of Enzymatic Activity

The group of antibiotics that has been produced industrially and widely studied are the β-lactam antibiotics. The techniques for metabolic engineering have been put to great use for this group of antibiotics for product improvement and product upscaling. The synthesis of penicillin and cephalosporin share part of the pathway. The sequential actions of three enzymes, namely ACV synthetase (ACVS), Isopenicillin N synthase (IPNS), and Acyltransferase(AAT), convert three amino acid precursors to penicillin. Cephalosporin C is synthesized by six enzymes in series by using the same amino acid precursors.

4.5.1.2 Manipulating Regulatory Genes and Shift to Biosynthesis from Semi-Synthetic Production

Natural products produced by plants, bacteria, and fungi have been a rich source of bioactive compounds for drug discovery and development. Natural products are also termed secondary metabolites or those not required for growth of the producing organisms. Unlike primary metabolites that are required for growth and are mostly the same across the spectrum of living organisms, secondary metabolites can vary widely from species to species and encompass a diverse array of complex chemical structures. Many of these compounds are structurally complex, containing multiple chiral centers and labile connectivities, which make them difficult to synthesize chemically. Biosynthesis and fermentative approaches are therefore important tools in the production and development of these compounds for pharmaceutical, agricultural, and related applications. In many organisms, pathway-specific regulators may be utilized to enhance production of the resulting natural product. Many gene clusters identified in *Streptomyces* encode a *Streptomyces* antibiotic regulatory protein (SARP), which has been shown to be a positive regulator of antibiotic production [19].

A regulatory gene *srm*R has been identified to have a role in the biosynthesis of spiramycin, a macrolide produced by *Streptomyces ambofaciens*. A mutant of *S. ambofaciens* derived by disrupting *srm*R was blocked in spiramycin production and was unable to complement any of the other mutants that accumulated different biosynthetic pathway intermediates in the medium. The product of *srm*R was identified as a transcriptional activator required for the initiation of transcription of the biosynthetic genes [20]. Introduction of the gene on a multicopy vector increased the final spiramycin titer from 100 to 500 µg/ml.

A pathway specific regulator *cca*R has also been identified recently in the *Streptomyces clavuligerus* cephamycin C gene cluster [21, 22]. Additional copies of this regulatory gene on a multicopy plasmid led to almost two-fold enhanced production of both cephamycin and clavulanic acid [23]. The disruption of the *cca*R gene led to the complete elimination of the biosynthesis of both compounds.

4.5.1.3   Enhanced Antibiotic Resistance

Many antibiotic producers have resistance mechanisms to protect themselves from the adverse effects of the antibiotics they produce. Mechanisms of cellular self-protection include [24]:

i.   drug inactivation by chemical modification
ii. target site modification, drug binding
iii. reduction of the intracellular concentration using an efflux pump system and/or a low permeability to the antibiotic.

Gene coding for antibiotic biosynthesis and resistance are often clustered, and their expression is physiologically interdependent. A molecular genetic basis for this interdependence has been demonstrated for the streptomycin producer *Streptomyces griseus*. A regulatory protein SmA controls the transcription of the shared promoter of the resistance gene and a pathway-specific activator, the latter in turn activates the biosynthetic genes [25]. Enhanced antibiotic resistance has been shown to be correlated to enhanced production for a number of antibiotic-producing species [26]. In *Streptomyces kanamyceticus* and *Streptomyces fradiae*, producers of kanamycin and neomycin, respectively, aminoglycoside 6'-N-acetyltransferase provides the host with resistance to these antibiotics by N-acetylation of the compound.

4.5.1.4   Increasing Product Selectivity

It is not necessary that one organism produces a single antibody; on the contrary, many antibiotic producers produce a number of antibiotics. The antibiotics produced might not be the end product, as they might be:

i.   intermediates in the biosynthetic pathway of any primary product.
ii. some by-products in branched pathways, if more products are the outcome of that pathway.
iii. others might be entirely different molecules from the intermediates or the byproducts.

Production of multiple byproducts might be undesirable as it becomes a burden for the energy resources of the organism, and also adds to the steps and cost involved in separation and downstream processing, thus affecting the overall yield, which is a great disadvantage for any industrial product. This phenomenon can be explained by the example of *S. pristinaespiralis*. This organism is used for the production of antibiotic pristinamycin. To produce a water-soluble derivative of this antibiotic, so that it can be given orally, a semisynthetic derivative was developed by chemical modification. The genes involved in the biosynthesis were identified and overexpressed which was under the control of a strong and constitutive promoter and elimination of undesirable products [32]. This was accomplished by using a site-specific integration plasmid vector to introduce a single additional copy of the gene.

4.5.1.5   Antibiotic Synthesis Utilizing Heterologous Strain

Heterologous expression of entire gene clusters for a complete pathway or segment of a pathway has been used for the production of non-native antibiotics in a number of organisms [27, 28] (Figure 4.6). Many reasons make it desirable to produce antibiotics in heterologous strains:

   i.   higher precursor flux.
   ii. better resistance to the end product.
   iii. fewer by products in the recipient organism.
   iv. ease to carry out all metabolic engineering in a heterologous host, for which better genetic tools are available or the physiological characteristics are better known.

Deacetoxycephalosporin C (DAOC) was produced in an industrial strain of *Penicillium chrysogenum* by transforming it with two hybrid genes, *cef*Dh and *cef*Eh, coding respectively for an isopenicillin N epimerase and DAOC synthase [29].

4.6   SCALING UP AN UPHILL TASK

Oxygen becomes a limiting factor in the industrial production of many antibiotics. In many filamentous microorganisms, oxygen transfer is a

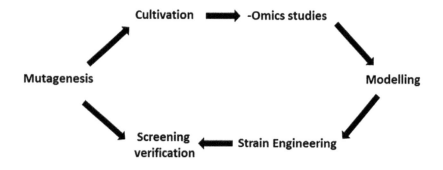

FIGURE 4.6   Production of heterologous strain.

major technical challenge in antibiotic production. This is because the dissolved oxygen level is often at a low concentration in large reactors. It becomes advantageous to increase the ability of microorganisms to utilize oxygen under such conditions. This has been done by introducing a bacterial hemoglobin into *Acremonium chrysogenum*, which is used for industrial production of cephalosporin [30]. This resulted in the production of higher levels of cephalosporin by the microorganism.

4.7 ENGINEERING PLANT METABOLIC PATHWAYS

Metabolic engineering of the plant metabolic pathway has fast emerged as a novel mechanism to obtain a desired product in large quantity. New approaches to manipulate plant metabolic pathways are continually emerging with the CRISPR-Cas9 genome engineering system as an important new genome-editing tool. Plants can be gene-edited with constructs that are composed exclusively of DNA sequence derived from the same or similar plant species that are called cis-genic plants. The cis-genic plants can be obtained through standard breeding practices and therefore may be more readily accepted by the public and regulatory bodies as opposed to transgenic plants. For example, as a mechanism to prevent blindness caused by vitamin A deficiency, a strain of rice was genetically engineered to express three heterologous genes that enabled the production of vitamin A.

4.8 PHYTOREMEDIATION

Plants are capable of removing environmental pollutants and convert them to nontoxic forms. Plants can be used to treat contaminated soil, sediments, and water, and the process is called phytoremediation, which is a clean technology. Added advantage of phytoremediation is that it causes less disruption to ecosystem as compared to physical, chemical, or microbial remediation. Apart from this, the plants are also able to stabilize contaminated soil and provide the conditions needed for microbial colonization of the rhizosphere, thereby facilitating symbiotic degradation and detoxification of pollutants. The disadvantage of phytoremediation is that it is time-consuming.

4.9   IMPACT OF NEW TOOLS

The biosynthetic pathways in case of many secondary metabolites is still elusive or unknown. This makes the use of metabolic engineering in microorganisms difficult. But despite this limitation, strain improvement is still required. For this, a new tool known as "Reverse Engineering" comes into play.

For most industrially important antibiotics, a large number of strains have been isolated in the course of strain improvement. "Reverse engineering" can possibly be combined with traditional strain improvement program to lead to a more "rational" approach of metabolic engineering. Another approach used is the comparison and analysis of genome sequence of wild-type and various high-producing strains. This analysis is of great use in understanding the mechanistic causes for the differences in the phenotype of wild-type and high-producing strains. The differences observed could be pin-pointed and used for strain modification and improvement. Such type of study has led to the discovery that high-producing strains of *Penicillium chrysogenum* have multiple copies of biosynthetic genes as well as higher transcript levels, which translate into enhanced product formation. Yet another mechanism would be to alter the substrate specificity to eliminate the synthesis of undesirable side products or manipulating the half saturation constant of the enzyme (Km) so that the precursors are channeled for biosynthesis of the product

4.10   METABOLIC ENGINEERING AND PROCESS ENGINEERING INTEGRATION

Integration of metabolic engineering with process engineering has paved the way for heterologous expression of proteins for secondary metabolite production when the original producer may not have the desired properties for large-scale cultivation in industrial reactors or may not have the sufficient fluxes of the precursors. This amalgamation may enable microorganisms to produce an antibiotic that is not normally produced by it. We have ushered into an era where possibility of expressing the entire pathway or a major portion of the pathway in a new host organism that exhibits more desirable characteristics can become a reality. Heterologous expression enables us to use a better host for the production. Heterologous expression can also facilitate the combination of one segment each from two different pathways. Each of the selection of particular segments of the pathway may be done on the basis of better reaction characteristics. The two segments may arise from the same or different species.

The biosynthetic machinery of secondary metabolites is temporally regulated. An integrated metabolic engineering approach takes factors such as temporal or spatial expression of genes involved in antibiotic production and the rate limiting enzymes into consideration. For example, the case of an enzyme that requires oxygen as a co-substrate being the rate limiting enzyme. Amplification of this enzyme may lead to an enhanced productivity in a laboratory investigation but not in a production reactor.

4.11   SUMMARY

The various technological advancements in the field of molecular biology have enabled the scientist to play at the genetic level of the microbes, bring about the desired modification/s, and make quantitative as well as qualitative changes in the desired product/s. Amalgamation of metabolic engineering with other traditional approaches for process improvement of many biochemicals, for example, antibiotic biosynthesis, is at an exciting stage where every day new inputs are coming in from research being conducted worldwide for further scope for improvement.

4.12   REVIEW QUESTIONS

1.   What is metabolic engineering?
2.   What is phytoremediation?
3.   Elaborate on microbial cell factory.
4.   How can metabolic engineering ease the burden of environmental pollution?
5.   What is cisgenic and why it is preferred over transgenic?
6.   How can common biosynthetic pathways be used for metabolic engineering?

KEYWORDS

- antibiotic biosynthesis
- metabolic engineering
- microbial cell factory
- phytoremediation

REFERENCES

1. Sauer, M., & Mattanovich, D., (2012). Construction of microbial cell factories for industrial bioprocesses. *J. Chem. Technol. Biotechnol.*, *87*(4), 445–450.
2. Qiu, J., Gao, F., Shen, G., et al., (2013). Metabolic engineering of the phenylpro-panoid pathway enhances the antioxidant capacity of *Saussurea involucrate*. *Plos One*, *8*(8), e70665.
3. Pandey, A., Misra, P., Bhambhani, S., et al., (2014). Expression of *Arabidopsis* MYB transcription factor, *AtMYB111*, in tobacco requires light to modulate flavonol con-tent. *Sci. Rep.*, *4*, 501.
4. Wilson, S. A., & Roberts, S. C., (2014). Metabolic engineering approaches for produc-tion of biochemicals in food and medicinal plants. *Curr. Opin. Biotechnol.*, *26*, 174–182.
5. Mandel, S. A., Amit, T., Weinreb, O., & Youdim, M. B. H., (2011). Understanding the broad-spectrum neuroprotective action profile of green tea polyphenols in aging and neurodegenerative diseases. *J. Alzheimer's Dis.*, *25*, 187–208. doi: 10.3233/JAD-2011-101803.
6. Cahoon, R. E., Lutke, W. K., Cameron, J. C., et al., (2015). Adaptive engineering of phytochelatin-based heavy metal tolerance. *J. Biol. Chem.*, *290*(28), 17321–17330.
7. Zaldivar, J., Nielsen, J., & Olsson, L., (2001). Fuel ethanol production from lignocel-lulose: a challenge for metabolic engineering and process integration. *Appl. Micro-biol. Biotechnol.*, *56*(1–2), 17–34.
8. Tatsis, E. C., & O'Connor, S. E., (2016). New developments in engineering plant metabolic pathways. *Curr. Opin. Biotechnol.*, *42*, 126–132.
9. Li, Y., Lin, Z., Huang, C., et al., (2015). Metabolic engineering of *Escherichia coli* using CRISPR-Cas9 meditated genome editing. *Metab. Eng.*, *31*, 13–21.
10. Cao, H. X., Wang, W., Hien, T. T., et al., (2016). The power of CRISPR-Cas9-in-duced genome editing to speed up plant breeding. *Int. J. Gen.*
11. Hater, K. F., (1990). The improving prospects for yield increase by genetic engineer-ing in antibiotic-producing Streptomycetes. *Biotechnology*, *8*, 115–121.
12. Hutchinson, C. R., Decker, H., Madduri, K., Otten, S. L., & Tang, L., (1993). Genetic control of polyketide biosynthesis in the genus Streptomyces. *Antonie Van Leeuwen-hoek.*, *64*, 165–176.
13. Hutchinson, C. R., (1994). Drug synthesis by genetically engineered microorgan-isms. *Biotechnology*, *12*, 375–380.

14. Lal, R., et al., (1996). Engineering antibiotic producers to overcome the limitations of classical strain improvement programs. *Critical Reviews in Microbiology*, *22*, 201–255.

15. Piepersberg, W., (1994). Pathway engineering in secondary metabolite-producing actinomycetes. *Crit. Rev. Biotechnol.*, *14*, 251–285.

16. Beppu, T., (1995). Signal transduction and secondary metabolism: Prospects for controlling productivity. *Trends Biotechnol.*, *13*, 264–269.

17. Horinouchi, S., & Beppu, T., (1994). A-factor as a microbial hormone that controls cellular differentiation and secondary metabolism in Streptomyces griseus. *Mol. Microbiol.*, *12*, 859–864.

18. Yamada, Y., Nihira, T., & Sakuda, S., (1997). Butyrolactone autoregulators, inducers of virginiamycin in Streptomyces virginiae: their structures, biosynthesis, receptor proteins, and induction of virginiamycin biosynthesis. In: *Biotechnology of Antibiotics 2 edn.* (Strohl, W. R., ed.), Marcel Dekker, New York, 63–79.

19. Chen, Y., Smanski, M., & Shen, B., (2010). Improvement of secondary metabolite production in *Streptomyces* by manipulating pathway regulation. *Appl. Microbiol. and Biotechnol.*, *86*, 19–25.

20. Geistlich, M., Losick, R., Turner, J. R., & Rao, R. N., (1992). Characterization of a novel regulatory gene governing the expression of a polyketide synthase gene in Streptomyces ambofaciens. *Mol. Microbiol.*, *6*, 2019–2029.

21. Waters, N. J., Barton, B., & Earl, A. J., (1994). Increasing yields of clavulanic acid – using DNA comprising a regulatory gene for clavulanic acid biosynthesis in host cells, e.g., *Streptomyces clavuligerus.*, *International Patent W0941826.*

22. Perez-Llarena, F. J., Liras, P., Rodriguez-Garcia, A., & Martin, J. F., (1997). A regulatory gene (ccaR) required for cephamycin and clavulanic acid production in Streptomyces clavuligerus: amplification results in overproduction of both beta-lactam compounds. *J. Bac.*, *179*, 2053–2059.

23. Cundliffe, E., (1989). How antibiotic-producing organisms avoid suicide. *Ann. Rev. Microbiol.*, *43*, 207–233.

24. Vujaklija, D., Ueda, K., Hong, S. K., Beppu, T., & Horinouchi, S., (1991). Identification of an A-factor-dependent promoter in the streptomycin biosynthetic gene cluster of Streptomyces griseus. *Mol. Gen. Genet.*, *229*, 119–128.

25. Demain, A. L., (1974). How do antibiotic-producing microorganisms avoid suicide? *Ann. N. Y. Acad. Sci.*, *235*, 601–612.

26. Malpartida, F., & Hopwood, D. A., (1984). Molecular cloning of the whole biosynthetic pathway of a Streptomyces antibiotic and its expression in a heterologous host. *Nature*, *309*, 462–464.

27. Smith, D. J., Burnham, M. K., Edwards, J., Earl, A. J., & Turner, G., (1990). Cloning and heterologous expression of the penicillin biosynthetic gene cluster from penicillium chrysogenum. *Biotechnology*, *8*, 39–41.

28. Cantwell, C., Beckmann, R., Whiteman, P., Queener, S. W., & Abraham, E. P., (1992). Isolation of deacetoxycephalosporin C from fermentation broths of Penicillium chrysogenum transformants: construction of a new fungal biosynthetic pathway. *Proc. R. Soc. Lond. B: Biol. Sci.*, *248*, 283–289.

29. DeModena, J. A., et al., (1993). The production of cephalosporin C by Acremonium chrysogenum is improved by the intracellular expression of a bacterial hemoglobin. *Biotechnology.*, *11*, 926–929.

CHAPTER 5

USE OF ENZYMES IN INDUSTRY

PRERNA PANDEY and ANJALI PRIYADARSHINI

CONTENTS

5.1   INTRODUCTION

Enzymes are the catalysts of biological systems and are used in industry and medicine because of their catalytic abilities, which ensure that they are substrate-specific and they remain unchanged throughout any reaction. Enzymes also keep costs down because they can be used in minimal quantities. They have been employed in the industries from creating lactose-free dairy products to fast-acting laundry detergents, to the textiles, foods, detergents, animals, and biofuels and also find application in therapeutics and healthcare industry. In medicine, they are useful because they are specific and do not cause side effects when administered to patients. Use of enzymes in industry and medicine is highly ethical, socially desirable and beneficial, economically efficient and represents an advance in modern technological processes.

5.2   ENZYMES IN THE FOOD INDUSTRY

The first industrial application of enzymes was recognized by the potential in fermentation. The conversion of starch to glucose employs amylase and *glucoamylase. β amylase* finds application in the conversion of starch to maltose and limits dextrin [1]. The enzyme transglutaminase is used in the processing of sausages, noodles, and yogurt, where cross-linking of proteins provides improved the elastic properties of the products [2].

α-amylases are employed in baking industry and increase quality and shelf-life [1]. As flour typically contains quantities of cellulose, glucans, and hemicelluloses like arabinoxylan and arabinogalactan apart from starch, other enzymes can be used. The use of xylanases decreases the water absorption, leading to more stable dough. Xylanases are used especially in wholemeal rye baking and dry crisps. The use of Proteinases to improve dough-handling properties; and lipases to strengthen gluten, which leads to more stable dough and better bread quality [3].

Pullulanases digest 1-6 linkages in carbohydrates and liberate 1,4 linkage containing straight chains, and help in debranching. Neopullulanases and amylopullulanases digest both 1,4 and 1,6 links (Figure 5.1) [4].

Glucose isomerases catalyze isomerization of glucose to fructose. This finds potential in slimming foods [5].

The major application of proteases in the dairy industry is for the manufacture of cheese. The use of calf rennin had been preferred in cheese-making due to its high specificity, but microbial proteases are

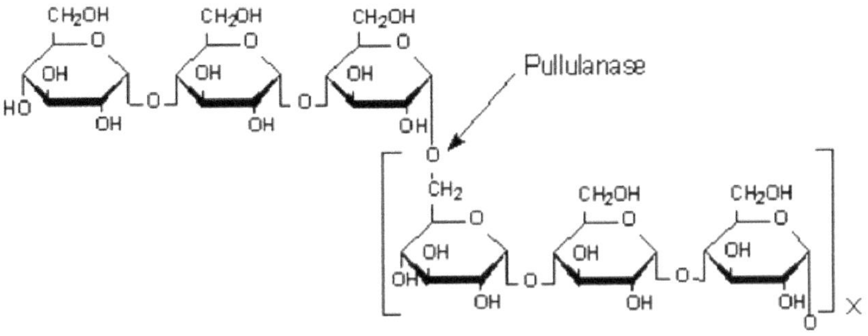

FIGURE 5.1 Chemical structure of pullulanase.

produced by GRAS (*Generally Recognized As Safe*) microorganisms like *Mucor miehei, Bacillus subtilis, Mucor pusillus Lindt,* and *Endothia parasitica.* Chymosin is used in cheese-making to coagulate milk protein. The use of β-galactosidase to breakdown milk-sugar lactose into glucose and galactose. This process is used for milk products targeted for lactose intolerant-consumers.

Bacterial glucose isomerase, fungal α-amylase, and glucoamylase are currently used to produce "high fructose corn syrup" from starch.

In the manufacture of fruit juices, the addition of pectinase, xylanase, and cellulase improves the formation of the juice from the fruit pulps. Pectinases and amylases are used in juice clarification to digest the pectin components of cell walls. α-amylase can be used to help starch hydrolysis, β-glucanases solve filtration problems caused by β-glucans present in malt, while papain controls haze during maturation, filtration, and storage. Naringase and limoninase remove compounds that impart bitter taste to citrus fruits.

In brewing, the enzyme α amylase also finds use as a liquifying adjunct-starchy cereals that are added to mash. Amyloglucosidases increase the glucose content and find use in the manufacture of light beer. They also increase soluble protein [6]. Proteases find use in improving yeast growth and improvement of malt. Pentosanases and xylanases hydrolyze pentosans of barley and wheat to improve extraction of beer and filtration. αacetolactate decarboxylases (ALDC) convert αacetolactate to acetoin to prevent diacetyl formation and decrease fermentation time. They also help in correcting beer taste. Pullulanases digest 1-6 linkages in carbohydrates to help in wort clarification [6].

5.3   ENZYMES FOR TEXTILES

Proteases, lipases, amylases, oxidases, peroxidases, and cellulases are added to detergents [7, 8]. Proteases are used to produce pharmaceuticals, foods, detergents, leather, silk, and agrochemical products. The first detergent containing a bacterial protease ("Biotex") was introduced by Novo Industry A/S (now Novozymes) in 1956 [9]. It contained an alcalase produced by *Bacillus licheniformis*. In 1994, Novo Nordisk introduced Lipolase™, the first commercial recombinant lipase for use in a detergent, by cloning the *Humicola lanuginose* lipase into the *A. oryzae* genome [9]. In 1995, Genencor International introduced two bacterial lipases, one from *Pseudomonas mendocina* (Lumafast™) and another from *Pseudomonas alcaligenes* (Lipomax™) [9, 10]. Amylases are used for de-sizing of textile fibers. Cellulase can modify the cellulosic fibers in a controlled and desired fashion so as to improve the quality of fabrics; cellulases are also utilized for digesting off the small fiber ends protruding from the fabric, resulting in a better finish [7, 8, 12].

Catalase is used to remove hydrogen peroxide, as hydrogen peroxides are used as bleaching agents. Laccase—a polyphenol oxidase from fungi—is a new candidate in this field. It is a copper-containing enzyme, which is oxidized by oxygen, and which in an oxidized state can oxidatively degrade many different types of molecules like dye pigments.

5.4   ENZYMES IN DETERGENTS

Dirt in the clothes could be proteins, starches, or lipids in nature. Vigorous mixing coupled with physical agitation may remove the stains but would shorten the life of the cloth. Enzyme detergents remove the proteins from clothes soiled with blood, milk, sweat, grass, etc. far more effectively than nonenzyme detergents. Cellulases are employed for loosening the fibers so as to remove the dirt easily and also it gives a finishing touch by digesting fine fibers during washing. At present, only proteases, amylases, cellulases, and lipases are commonly used in the detergent industry [10–12].

Lipases are added to detergents [12] such as household and industrial laundry and in household dishwashers, where their function is in the removal of fatty residues and cleaning clogged drains [11] and industrial laundry and in household dishwashers, where their function is in the removal of fatty residues and cleaning clogged drains [11, 12].

Proteases are the most widely used enzymes in the detergent industry to remove protein stains such as grass, blood, egg and human sweat, which have a tendency to adhere strongly to textile fibers. Amylases are used to remove residues of starch-based foods like potatoes, spaghetti, custards, gravies, and chocolate. Lipases decompose fatty material. Lipase is capable of removing fatty stains such as fats, butter, salad oil, sauces and the tough stains on collars and cuffs. Cellulases modify the structure of cellulose fiber on cotton and cotton blends. When it is added to a detergent, it results in color brightening, softening, and gives a nice finishing touch when added to the fibers [12–14].

5.5   ENZYMES USED IN THE PAPER INDUSTRY

Lipases, xylanases, and laccases find use in removing pitch (hydrophobic components of wood). Lipases can digest the sticky lipids of wood that cause issues in printing. Xylanases digest xylan of wood, thus finding potential in bleaching instead of the harsh chemicals. Mannases digest glucomannin and enhance paper brightness. Laccases help in bleaching. Cellulases hydrolyze cellulose and create weak spots in paper, enhancing flexibility and deinking. Amylases reduce the viscosity as they cleave starch molecules [12–14].

5.6   ENZYMES USED IN ANIMAL HUSBANDRY

Feed enzymes can increase the digestibility of nutrients, leading to greater efficiency in feed utilization [15]. Also, they can degrade unacceptable components in feed, which are otherwise harmful or with no value.

Diets based on cereals such as barley, rye, and wheat are higher in non-starch polysaccharides (NSPs), which can decrease the intestinal methane production when supplemented with NSP enzymes. Furthermore, proteases can substantially reduce the amount of non-protein nitrogen supplement in diets of animals, thereby reducing the excretion of urea into the environment [15].

The first commercial success was the addition of β-glucanase into barley-based feed diets. Barley contains β-glucan, which causes high viscosity in the chicken gut. The net effect of enzyme usage in the feed has been increased animal weight gain with the same amount of barley, resulting in increased feed conversion ratio. Proteases (Subtilisin) help to produce amino acids to enhance nutrition.

Addition of xylanase to feed to viscous diets is documented. Phytase, which is a phosphoesterase, liberates phosphorous from the phytic acid (in plant-based feed materials).

5.7   ENZYMES USED IN THE LEATHER INDUSTRY

Traditionally, tanneries apply enzymes in the bate step to perform deep cleaning of the hide [16, 17]. However, enzymes were also used in the hair removal process at the beginning of the last century before the development of chemical processes for hair removal. The main enzymes that are as follows:

- Proteases because they hydrolyze the protein fraction of dermatan sulfate, making the collagen more accessible to water and reducing the attachment of the basal layer. In addition, they act in the removal of globular proteins and help in reducing the processing time.
- Lipases, which hydrolyze fats, oils, and greases present in the hypoderm.

Keratinases hydrolyze the keratin of hair and epidermis and break down the disulfide bonds of this molecule.

Recently, a novel keratinase from *Bacillus tequilensis* strain Q7 with promising potential for the leather bating process was extensively investigated [18].

The use of pepsin on chrome tanned skins and hides have been reported. It also acts as a good degreasing agent.

5.8 BIOSENSORS

Biosensors are hybrid analytical devices that amplify signals generated from the specific interaction between a receptor and the analyte, through a biochemical mechanism. Biosensors use tissues, whole cells, artificial membranes or cell components like proteins or nucleic acids as receptors, coupled to a physicochemical signal transducer [19].

Biosensors consist of a receptor system, in which a biological molecule under study interacts specifically with a given analyte, and a coupled physicochemical transducer that amplifies the signal resulting from such interaction. The five principal transducer classes are [19]:

- electrochemical,
- optical,
- thermometric,
- piezoelectric, and
- magnetic.

The purpose of these constructs is the identification, quantification, and eventual screening of specific molecules, as present in complex mixtures from moderate to very low concentrations. Therefore, biosensors have utility in analytical research, clinical diagnosis, the food and pharmaceutical industries, environmental control, and process monitoring.

A common example is the blood glucose biosensor, which uses the enzyme glucose oxidase to break down blood glucose. The basic concept of the glucose biosensor is based on the fact that the immobilized glucose oxidase catalyzes the oxidation of glucose by molecular oxygen producing gluconic acid and hydrogen peroxide [20, 21]. The enzyme requires a redox cofactor—flavin adenine dinucleotide (FAD). FAD works as the initial electron acceptor and is reduced to $FADH_2$ [20–22].

$$Glucose + GOx-FAD+ \rightarrow Glucolactone + GOx-FADH_2$$

The cofactor is regenerated by reacting with oxygen, leading to the formation of hydrogen peroxides.

$$GOx-FADH_2 + O_2 \rightarrow GOx-FAD+H_2O_2$$

Hydrogen peroxide is oxidized at a catalytic, classically platinum (Pt) anode. The electrode easily recognizes the number of electron transfers, and this electron flow is proportional to the number of glucose molecules present in blood.

$$H_2O_2 \rightarrow 2H^+ + O_2 + 2e$$

5.8.1 ELECTROCHEMICAL SENSING OF GLUCOSE

Three general strategies are used for the electrochemical sensing of glucose: by measuring oxygen consumption, by measuring the amount of hydrogen peroxide produced by the enzyme reaction, or by using a diffusible or immobilized mediator to transfer the electrons from the enzyme to the electrode [20–22].

Noninvasive glucose analysis has been developed by employing optical or transdermal approaches.

The launch of ExacTech in 1987 accelerated the portable glucose biosensors and has achieved the most significant commercial success. Subsequently, many different devices have been introduced in the global market [23].

5.8.2 NANOBIOSENSORS

These are basically the sensors made up of nanomaterials. These are the materials that have one of their dimensions between 1 and 100 nanometers [24]. Some of such materials that are widely employed include nanotubes, nanowires, nanorods, nanoparticles, and thin films made up of nanocrystalline matter. These can be as diverse as using amperometric devices for enzymatic detection of glucose to using quantum dots as fluorescence agents for the detection of binding and even using bio-conjugated nanomaterials for specific biomolecular detection. These include colloidal nanoparticles that can be used to conjugate with antibodies for immunosensing and immunolabeling applications. Further, metal-based nanoparticles are very excellent materials for electronic and optical applications and can be efficiently used for the detection of nucleic acid sequences through the exploitation of their optoelectronic properties [24, 25].

5.8.3   ACOUSTIC WAVE BIOSENSORS

These have been developed employing particles of gold, platinum, cadmium sulfide, and titanium dioxide to amplify the sensing responses so as to improve the overall preciseness of the limits of biodetection.

Special devices such as superconducting quantum interference devices (SQUID) have been used for rapid detection of biological targets using the super paramagnetic nature of magnetic nanoparticles. These devices are used to screen the specific antigens from the mixtures by using antibodies bound to magnetic nanoparticles [26].

An application of biosensors in the detection of various diseases may be illustrated. A nanobiosensor (based on magnetic nanoparticles) has been developed for rapid screening of telomerase activity in biological samples. The technique makes use of nanoparticles which, upon annealing to telomerase-synthesized telomeric repeats (TTAGGG), change their magnetic state. This helps to monitor telomerase activity and hence, detection of cancer [26, 27].

Certain biosensors can be used in the food industry. Lactic acid and lactate fermentation play important role in food and beverages production, control and quality. Lactate biosensors can detect the levels of lactate in various products. The assay of glucose in meat surface can reveal its shelf-life. The assay of glutamate and lactate (using lactate oxidase) especially in the cheese industry offers information on the quality of cheese [25–27].

5.9   CLINICAL USES OF ENZYMES

Preparation of chiral medicines, i.e., the synthesis of complex chiral pharmaceutical intermediates efficiently and economically, is one of the most important applications in biocatalysis. Esterases, lipases, proteases, and KREDs (ketoreductases) are widely applied in the preparation of chiral alcohols, carboxylic acids, amines or epoxides, among others.

Pure chiral amines are crucial building blocks in the synthesis of some pharmaceutical drugs, agrochemicals, and other chemical compounds. Amine transaminases (ATAs) are important objects of study. The biocatalytic activity of these enzymes that holds great potential in the study of these enzymes and target specific experiments [28].

Codexis, in cooperation with Pfizer, produced 2-methyl pentanol, an important intermediate for the manufacture of pharmaceuticals and liquid crystals [28]. Sitagliptin, the active ingredient in Januvia which is a leading drug for type 2 diabetes, was replaced by a new biocatalytic process. Several rounds of directed evolution were applied to create an engineered amine transaminase with a 40,000-fold increase in activity [29]

An enzymatic process for the production of montelukast (Singulair) and silopenem was developed by Liang, et al. [30]. They also developed an improved LovD enzyme (an acyltransferase) for improved conversion of the cholesterol-lowering agent, lovastatin, to simvastatin.

5.10   ENZYMES IN COSMETICS

Oxidases and peroxidases are used in dyeing of hair. Protein disulfide isomerases and transglutaminases find use in hair-based applications at saloons. Papain and proteases find use in facial peels. Glucose oxidases and aminoglucosidases are used in toothpastes and mouthwashes. Other applications include catalase in skin protection, laccase in hair dye, lipase to prevent treat skin rash or diaper rash, and endoglycosidase and papain in toothpaste and mouthwash to whiten teeth and remove plaque and odor-causing deposits on teeth and gum tissue [31].

Contact lens cleaning employs hydrogen peroxide for their disinfections. The residual hydrogen peroxide after disinfections can be removed by a heme-containing catalase enzyme, which degrades hydrogen peroxide. Proteinase- and lipase-containing enzymes can also be used [31].

5.11   ENZYMES IN RECOMBINANT DNA TECHNOLOGY

The technology is the creation of recombinant DNA molecule that is a recombinant molecule where the vector is joined with a natural or synthetic DNA segment of interest that is capable of replication in a host [32,

33]. It employs digesting DNA using restriction endonucleases and subsequent joining or ligation using DNA ligases.

The following are the enzymes used:

5.11.1 NUCLEASES

These digest DNA by hydrolyzing the phosphodiester linkages on the DNA backbone. They are of two types: exonucleases and endonucleases.

5.11.1.1 Exonucleases

These digest the DNA and the bonds at the terminal ends of a DNA strand. Example: Bal31 is isolated from a marine bacterium *Alteromonas espejiana*. It is a Ca^{2+}-dependent enzyme that degrades the nucleotides from both the strands of dsDNA molecule. Another enzyme is exonuclease III that digests only one strand of the dsDNA molecule. It removes the nucleotide from the 3′ terminus of the strand, thus leaving protruding 5′ overhangs. Exonuclease III is used for generating single-stranded templates [32, 33].

5.11.1.2 Endonucleases

These digest a DNA strand at internal sites. S1 nuclease is an endonuclease isolated from *Aspergillus oryzae*. It cleaves only single-stranded DNA (ssDNA) and single stranded nicks in dsDNA molecules. DNase I isolated from cow's pancreas is a nonspecific enzymes. It is able to cleave both ssDNA and dsDNA. It can cleave any of the internal phosphodiester bonds; thus, prolonged digestion of DNA with DNase I results in its complete chewing, leaving only a mixture of mononucleotides [32, 33].

5.11.1.3 Restriction Endonucleases

Restriction endonucleases are a type of endonucleases that cleave of DNA at "particular sites." These restriction sites are usually palindromic. These enzymes may either yield sticky ends or blunt ends. In a blunt-ended molecule, both strands terminate in a base pair. *Cohesive ends* or *sticky ends*. They are most often created by enzymes when they cut DNA in a staggered fashion. Very often, they cut the two DNA strands few base pairs from each other, creating a 5' overhang in one molecule and a complementary 3' overhang in the other. These ends are called cohesive since they are easily joined back together by a ligase. Also, since different restriction endonucleases usually create different overhangs, it is possible to cut a piece of DNA with two different enzymes and then join it with another molecule [32, 33].

5.11.2   LIGASES

Ligases are enzymes that join the nucleic acid molecules together. These nucleic acids can either be DNA or RNA, and the enzymes are thus called DNA ligase and RNA ligase, respectively. DNA ligase catalyzes the formation of a phosphodiester bond between the 5' phosphate of one strand and the 3' hydroxyl group of another. In nature, the function of DNA ligase is to repair single strand breaks (discontinuities) that arise as a result of DNA replication and/or recombination. In recombinant DNA technology, ligases catalyze the joining of DNA of interest called as "insert," with the vector molecule and the reaction is known as ligation. For molecular cloning, the most commonly used DNA ligase is obtained from bacteriophage T4. T4 DNA ligase requires ATP as cofactor and Mg2+ions for its activity. It can perform blunt end as well as sticky end ligations [32, 33].

5.11.3   POLYMERASES

These enzymes catalyze the addition of nucleotides to a preexisting template strand, i.e., DNA synthesis. Several different types of polymerases have specific uses. *E. coli* DNA polymerase I (PolI) is an enzyme that has both DNA polymerase as well as DNA nuclease activity. This enzyme binds to the "nick" region (region of a double-stranded DNA where one or more nucleotides of one strand are missing, making it single stranded). If the *E. coli* Pol I holoenzyme is treated with a mild protease, it results in the formation of two fragments. A larger fragment retaining both 5'-3' polymerase and 3'-5' exonuclease activities, while the smaller one has only the 5'-3' exonuclease activity. The larger fragment is known as "Klenow fragment." This Klenow fragment can synthesize the new DNA strand complementary to the template but cannot degrade the existing strand. Klenow fragment is predominantly used in DNA sequencing. Other uses in recombinant DNA technology where Klenow fragment is used are synthesis of dsDNA from single stranded template, filling of 5' overhangs created by restriction enzymes to create blunt ends, and digestion of protruding 3' overhangs to produce blunt ends [32, 33].

5.11.3.1   Thermostable DNA Polymerases

These are a class of DNA polymerases that remain functional even at high temperatures. Taq polymerase isolated from the bacterium *Thermus aquaticus* is a commonly employed enzyme. Pfu polymerase from *Pyrococcus furiosus* is another enzyme. These enzymes are employed in polymerase chain reaction [32, 33].

5.11.3.2   Reverse transcriptase (RT)

RT is a RNA-dependent DNA polymerase found in certain retroviruses. RT uses mRNA template instead of DNA for synthesizing new DNA strand. The complementary DNA strand formed on the mRNA template is called complementary DNA (cDNA). RT also shows RNAse activity that degrades the RNA molecule from a DNA-RNA hybrid. Formation of a double stranded cDNA from the mRNA molecule using RT finds applications in genetic engineering. The cDNA thus formed from any mRNA can be cloned in an expression vector [32, 33].

5.11.4   *ALKALINE PHOSPHATASE (AP)*

This group of enzymes removes the phosphate group from 5' terminus of the DNA molecule. It is active at alkaline pH, and hence the name "alkaline phosphatase." Commercially, it is obtained from three major sources, viz., *E. coli* (bacteria), calf intestine, and arctic shrimp. Removal of 5' phosphate prevents self-annealing of the digested vector and increases the possibility of ligating with the insert DNA fragment in the presence of ligase. Also, radiolabeled DNA probes are prepared by initially removing the 5' PO_3^{2-} by AP treatment, followed by polynucleotide kinase treatment in the presence of radioactive phosphate [32, 33].

5.11.5   POLYNUCLEOTIDE KINASE (PNK)

PNK catalyzes the transfer of a phosphate group from ATP to the 5' terminus of the DNA molecule. This enzyme is obtained from E. coli infected with T4 phage [32, 33].

Terminal transferase: This group of enzymes catalyzes the addition of one or more deoxyribonucleotides to the 3' terminus of the DNA molecule. The enzyme is used for labeling 3' ends of DNA. Also, it can be used for adding complementary homopolymeric tails to DNA molecules [32, 33].

5.12   PEPTIDE ANTIBIOTICS

Due to the threat of bacterial resistance, the quest for new antimicrobial agents or drug targets prompted research in short peptides called antimicrobial peptides (AMPs). The discovery of AMPs dates back to 1939, when Dubos extracted an antimicrobial agent from a soil *Bacillus* strain. This extract was demonstrated to protect mice from pneumococci infection [34].

Frog skin has been used for medicinal purposes for centuries and is still used today in South American countries. In 1962, Kiss and Michl noted the presence of antimicrobial and hemolytic peptides in the skin secretions of *Bombina variegata*, and this led to the isolation of a 24-amino-acid antimicrobial peptide named "bombinin" [35].

Antimicrobial peptides are a unique and diverse group of molecules, which are divided into subgroups on the basis of their amino acid composition and structure. Peptide antibiotics are produced by bacterial, mammalian, insect, and plant organisms in defense against invasive microbial pathogens. Through evolution, host organisms developed varieties of AMPs that protect them against a large variety of invading pathogens including both gram-negative and gram-positive bacteria [34, 35].

AMPs are amphipathic molecules with hydrophobic and hydrophilic groups in distinct patches on the molecular surface. Due to the frequent presence of lysine and arginine residues in their amino acid sequence, they usually possess a net positive charge at physiological pH   (Anionic peptides have been reported in toads.) [35].

AMPs are usually gene-coded and can be constitutively expressed or induced to fend off invading pathogens. A few examples are mentioned below.

5.12.1   INSECT

A few AMPs in insects include Drosomycin in Fresh fly, Sapecin and, Drosocin in *Drosophila melanogaster*; Trachyplesin I in crab *Tachypleus tridentatus*; PR-39 in pig.

5.12.2   PLANT PEPTIDES

Thionins were the first antimicrobial peptides to be isolated from plants. Other antimicrobial peptides were isolated that were found to be structurally related to insect and mammal defensins and have been named "plant defensins" [36].

5.12.3   HUMANS

They have been identified in a variety of exposed tissues or surfaces such as skin, eyes, ears, mouth, airways, lung, intestines, and the urinary tract [37]. Few examples include human cathelicidin LL-37 is detected in the skin of new born infants, psoriasin (S100A7), RNase 7, and hBD-3 are differentially expressed in healthy human skin; psoriasin is upregulated upon breakage of the skin barrier. In addition, β-defensins are expressed in human middle ear epithelial cells. Drosomycin-like defensin (DLD) is produced in human oral epithelial cells as part of host defense against fungal infection. Defensins have been reported as natural peptide antibiotics of human neutrophils [37].

Most AMPs are produced by specific cells at all times, while the production of some AMPs is inducible.

Antimicrobial peptides enhance the ion permeability of lipid bilayers. Ion channel or pore formation and the destruction of the electrochemical gradient across the cell membrane are thought to be the main killing mechanism of the vast majority of AMPs. In the case of certain AMPs, the possibility of the existence of intracellular targets has been implied (Apidaecins point at components of the protein synthesis machinery as likely candidates) [38]. AMPs are also known to enhance the activities of antibiotics through synergistic effects [38].

Because AMPs are composed of amino acids, they have potential to modify the structure (including library construction and screening) and immobilize AMPs on surfaces. It is possible to make fully synthetic peptides by either chemical synthesis or by using recombinant expression systems [38].

Ali Adem Bahar et al. [39] have reviewed the various requirements of AMPs, for example, many of the AMPs are cationic, targeting bacterial cell membranes, and cause disintegration of the lipid bilayer structure or AMPs may affect internal pathways like replication [39].

5.12.4 SYNTHESIS OF AMPS

There are 2 classes: ribosomally synthesized peptides, such as gramicidins, polymyxins, bacitracins, glycopeptides, etc., and ribosomally synthesized (natural) peptides.

Peptide antibiotics possess a variety of chemical structures like cyclic di- and oligopeptides. depsipeptides, linear peptides with repeating sequences of Land D-amino acids, and substituted peptides containing non-peptide components [39].

Nonribosomally synthesized peptides are elucidated as peptides elaborated in bacteria, fungi, and streptomycetes that contain two or more moieties derived from amino acids. For example, bacitracin, gramicidin S, and polymyxin B.

Multiple-carrier thiotemplate mechanism has been implicated in the biosynthesis of peptides. In this template-driven assembly, a series of very large multifunctional peptide synthetases, with a modular arrangement, perform the peptide synthesis in an ordered fashion [39]. Each module contains the basic ability to recognize a residue, activate it, modify it as necessary, and add it to the growing peptide chain. The minimal module is capable of activating one amino acid or hydroxyacid residue, stabilizing the activated residue as a thioester, and polymerizing it in its correct sequence to the previously added residue with the aid of a covalently attached cofactor. This basic mechanism can result in a great chemical variety of peptide products containing hydroxy-, L-, D-, or unusual amino acids, which can be further modified by N –methylation, acylation, glycosylation, or heterocyclic ring formation.

Ribosomally synthesized peptides: Antimicrobial, ribosomally synthesized peptides have been recognized as an important part of innate immunity across a spectrum of organisms that have been discussed above. However, little sequence homology has been revealed when subjected to analysis suggesting that each peptide has evolved (probably convergently) to act optimally in the environment in which it is produced and against local microorganisms [40].

Amino acid and nucleic acid analogs have also been employed to distinguish between antibiotic peptide and protein synthesis. The results obtained with inhibitors of protein, RNA, and DNA synthesis indicated that there is no direct involvement of DNA or RNA in peptide antibiotic synthesis [40].

In vivo studies have also suggested that antibiotic formation is catalyzed by enzymes with relatively broad specificities. Amino acid substitutions in peptide antibiotics have been reported by several investigators. Tyrocidines, actinomycins, quinomycins, and many other peptide antibiotics synthesized by an organism normally differ, respectively, by single amino acid substitutions. It appears more likely that amino acid substitutions in antibiotic peptides are probably not the result of different templates arising through gene mutations, but rather depend on physiological and environmental factors.

Gevers et al. [41] advanced the interesting hypothesis that antibiotic peptide synthesis may be analogous to fatty acid elongation that occurs in a polyenzyme complex (fatty acid synthetase) [41]. On the basis of this similarity, it was suggested that pantotheine may play a role in peptide antibiotic synthesis. It is suggested that phosphopantotheine acts like an arm to accept the growing peptide chain after each successive peptide bond is formed and to transfer the peptide to the next thioester-linked amino acid in sequence [41].

Imino acids, N-methylamino acids, n-amino acids, and D-amino acids are unique to peptide antibiotics. In most cases, in vivo studies have revealed that the L-amino acid, but not the D-isomer, is the precursor of the D-amino acid in the antibiotic. Yukioka and Winnick [42] observed that both r- and L-leucine were utilized for malformin synthesis in vitro [42].

Research revealed they are synthesized by soluble fractions derived from bacteria and are not affected by inhibitors of protein synthesis.

Research indicates AMPs with very similar structures can have drastically different mechanisms of action and the range of targeted cells. Hence, the following pointers would assist the design of synthetic AMPs.

Apart from the length of the AMP, the net charge either cationic or anionic plays a key role in the design. Alteration in the charge causes drastic changes in its activity. The incorporation of D-amino acids in the design of the structure shows promise as they are not sensitive to protease treatment.

Almost half proportion of amino acids in the primary sequence of natural AMPs are hydrophobic residues. Research points to increase in hydrophobicity on the positively charged side of an AMP below a certain limit can enhance its antimicrobial activity, while decreasing hydrophobicity can reduce antimicrobial activity. Hence, an optimum value must be kept in mind for the design [41–43].

Amphipathicity is an integral point in the design as the AMPs interact with cell membranes.

Naturally, AMPs are subjected to different post-translational modifications such as phosphorylation, addition of d-amino acids, methylation, amidation, glycosylation, formation of disulfide linkage, and proteolytic cleavage.

Research shows that while designing synthetic AMPs, the use of special modification like amidation, i.e., addition of amide groups especially at the C-terminal ends, has profound effects on enhancing its uptake and activity.

Also, the incorporation of unnatural amino acids such as β-didehydrophenylalanine enhances its abilities. A rare amino acid residue component of teixobactin, enduracididine, is only known to occur in a small number of natural products that also possess promising antibiotic activity has been reported by Atkinson et al. [43].

The use of artificial intelligence and computers in the design of AMPs has been of interest. Jenssen et al. [44] has employed QSAR modeling and computer-aided design of antimicrobial peptides [44]. Using peptide array technology together with artificial neural networks, several interesting molecules were obtained by Cherkasov et al. [45].

5.13 CATALYTIC ANTIBODIES

The immune response in an organism employs the production and action of proteins called immunoglobulins. Antibodies molecules are produced by the immune system to eliminate foreign particles called antigens. Antigens include microbes as well as certain small chemical molecules called haptens that can elicit the production of antibodies.

Linus Pauling postulated his theory that enzymes achieve catalysis because of their complementarity to the transition state for the reaction being catalyzed. For catalysis to occur, the energy of the transition state has to be lowered. Formation of this transition state geometry is energetically unfavorable in the absence of enzyme [46].

In 1969, W. P. Jencks proposed that an enzyme can be produced by making an antibody against the transition state analog of the reaction concerned. If an antibody binds to a transition-state molecule, it may be expected to catalyze a corresponding chemical reaction by forcing substrates into transition-state geometry. The first abzymes produced accelerated the hydrolysis of esters about 10^3-fold.

Catalytic antibodies can enhance a couple of chemical and metabolic reactions in the body by binding a chemical group, resembling the transition state of a given reaction [46]. An important point to notice is that ordinary antibodies do not chemically alter an antigen they bind to and neutralize.

Catalytic antibodies occur naturally in healthy individuals where they may form part of the innate immune system but are preferentially found in those with autoimmune disease. They can also be artificially engineered or elicited by immunizations [46].

They are also called abzymes (ab from antibody). The first natural catalytic antibody, now termed abzyme, which hydrolyzes intestinal vasoactive peptide, was discovered by Paul et al. [47]. Subsequently, other abzymes able to hydrolyze proteins, DNA, RNA, or polysaccharides have been found in the sera of patients with autoimmune and viral pathologies.

Because antibodies are highly specific and diverse, it has been recorded that high numbers are produced and their numbers enhanced by somatic mutation. These properties enable the production of tailor-made antibodies.

The use of hybridoma technology has opened up new vistas in the field of antibody engineering. The production of specific monoclonal antibodies against transition state analogs have been employed for several purposes.

5.13.1   OVERVIEW

5.13.1.1   Natural Activity

Antibodies-IgM and IgG show proteolytic activities that have protective roles. The cleavage of HIV gp120 by IgM of uninfected individuals has been reported by Paul et al. [48].

5.13.1.2   Production

The initial stage is the design and synthesis of a transition state analog of the desired reaction. The approach for enzymatic transition state analysis involves the selection of an enzyme target, and synthesis of the isotopically labeled substrates. This is followed by measurement of intrinsic kinetic isotope effects and usage of wave function calculations and application of synthetic organic chemistry to the synthesis of the mimics and subsequent testing of the analogs against the enzyme.

The generation of antibodies employs monoclonal antibody production. But due to the low number of desired clones obtained, various modifications have been developed. The employment of combinatorial variable region cloning in phages has been established [49].

The use of an idiotypic pathway to produce an abzyme is another approach. This method employs the generation an antibody that is complementary to the active site of the desired enzyme and then raising a second antibody to the variable region of the first antibody [50].

Several reactions that have been catalyzed by monoclonal antibody abzymes include:

Mimicking the transition state for the hydrolysis of methyl p-nitrophenyl carbonate with the phosphonate ester. The approach was extended to carbonic esters. The hydrolysis of cocaine's ester prevents its effect; Landry et al. [51] generated a monoclonal abzyme that had esterase activity. This shows potentials in the treatment of human diseases [51].

Additionally, abzymes showing amide hydrolysis, cyclisation (Diel-Alder reaction), formation of amide bond, and decarboxylation to name a few have been reported [52].

Apart from monoclonal abzymes, polyclonal abzymes show an added advantage of their ease in production and potential in technological processes.

Natural catalytic antibodies were first found in blood of asthmatic patients by Paul et al. (1989). A small fraction of the antibodies was able not only to bind VIP but also to hydrolyze it. This hydrolytic activity was highly specific with respect to VIP; it was not observed with other peptides and proteins. The first polyclonal abzymes generated were obtained by immunizing sheep with 4-nitrophenyl 4'-(carboxymethyl)phenyl hydrogen phosphate. The authors reported better hydrolysis of aryl carbonate compared to monoclonal abzymes [53].

5.13.2 POTENTIALS

The anti-cocaine abzyme generated by Landry et al. [51] has been discussed in the preceeding paragraph [51]. Similarly, abzymes that degrade nicotine have been reported by Dickerson et al. [54].

ADEPT (Antibody-Directed Enzyme Prodrug Therapy): An antibody designed/developed against a tumor marker linked to an enzyme is injected to the blood, resulting in selective binding of the enzyme in the tumor. This is followed by the administration of a prodrug that is activated in the vicinity of the tumor cell. The usage of radioimmuno-congugates where toxins are coupled with antibody molecules have also been reported [55, 56].

These principles can be used for ADAPT (Antibody-Directed Abzyme Enzyme Prodrug Therapy). Prodrug activation by the catalytic antibodies that have high specificity is being established [55, 56].

Monoclonal antibody 6D9 was the first for activating ester prodrug into chloramphenicol. The catalytic antibody 38C2 with aldolase activity was successfully tested to deliver insulin in prodrug from in vivo [57].

The abzymes have shown success in destruction of tumor tissue as an instance given below. Wentworth, et al. [56] reported the production of a catalytic antibody that can hydrolyze the carbamate prodrug 4-[N,N-bis(2-chloroethyl)]aminophenyl-N-[(1S)-(1,3- dicarboxy)propyl]carbamate to generate the corresponding cytotoxic nitrogen mustard [56]. In vitro studies with this abzyme, EA11-D7, and prodrug lead to a marked reduction in viability of cultured human colonic carcinoma (LoVo) cells.

Antibodies have been generated as reported by Zorrilla (2006) to degrade Ghrelin, which is involved in weight gain; this could prove to be a potential shield against weight gain [58].

Paul et al. [59] reported the results of efforts to strengthen and direct the natural nucleophilic activity of antibodies (Abs) for the purpose of specific cleavage of the human immunodeficiency virus-1 coat protein gp120. (Specific HIV gp120-cleaving antibodies induced by covalently reactive analog of gp120) [59].

5.14   NUCLEIC ACIDS AS CATALYSTS

The discovery of natural RNA catalysts has prompted chemical biologists to pursue artificial nucleic acids that have catalytic activities. Such artificial nucleic acid enzymes may comprise either RNA (ribozymes) or DNA (deoxyribozymes) [60]. In 1982, Kruger et al. reported natural catalytic activity by RNA, while "deoxyribozyme" was first used in 1994, when Breaker and Joyce reported the first artificial DNA catalyst [60, 61].

The known natural ribozymes catalyze phosphodiester cleavage or ligation, with the exception of the ribosome—made of both RNA and protein—that catalyzes peptide bond formation. Many artificial ribozymes also catalyze phosphodiester exchange reactions (RNA/DNA cleavage or ligation), although a growing number of ribozymes catalyze other reactions [61]. Artificial ribozymes for phosphodiester cleavage or ligation have been emphasized in part because nucleic acid catalysts can readily bind via Watson–Crick base pairs to oligonucleotide substrates. Within the ribosome, ribozymes function as part of the large subunit ribosomal RNA to link amino acids during protein synthesis. They also participate in a variety of RNA processing reactions, including RNA splicing, viral replication, and transfer RNA biosynthesis.

5.14.1   GROUP I INTRON

The Group I Intron from the ciliate Tetrahymena pre-rRNA was the first ribozyme to be discovered [60, 61]. Group I introns catalyze the self-excision of intronic sequences through two separate phosphoryl-transfer reactions. In the first step, the 5'-exon–intron junction is cleaved following attack by the 3'-oxygen of an exogenous guanosine cofactor. This reaction adds the guanosine to the intron and leaves a 3'-hydroxy terminus on the 5'-exon. A structural rearrangement then positions the 3'-intron–exon junction for attack by the newly created terminal 3'-oxygen of the 5'-exon.

5.14.2   GROUP II INTRONS

The Group II Introns are the second largest of the naturally occurring ribozymes and are an extremely diverse family with a strong relationship to the eukaryotic splicing machinery. Unlike the group I introns, in group II introns, the initial nucleophilic attack on the 5'-exon junction is performed by either an endogenous 2'-OH group (as opposed to attack by the 3'-OH attack of an exogenous nucleotide in group I introns) or a water molecule, leading to a lariat or linear intermediate, respectively [62]. The second reaction, as for the group I intron, uses the 3'-OH leaving group of step one to attack the 3'-exon junction, excising either a circularized or linear intron. Both phosphoryl-transfer reactions are highly reversible, and the intron can be designed to target any sequence. Group II introns have the additional ability to use DNA as a natural substrate for the reverse reaction. Some group II introns contain open reading frames that code for a reverse transcriptase "maturase." Excised group II introns can use the highly reversible nature of their reactivity to invade duplex DNA in a process called "retrohoming" [62].

5.14.3   RNase P

RNase P catalyzes the cleavage of a specific phosphodiester bond during the 5'-maturation of tRNA, cleaving off a "leader" sequence to yield 5'-phosphate-terminated tRNA. RNase P catalyzes the attack of a phosphodiester bond in the 5'-leader sequence of pre-tRNA by an exogenous water or hydroxide nucleophile. Distinct from the other nucleolytic ribozymes, the products of the RNase P reaction are a tRNA with a 5'-monophosphate and a leader sequence with a 3'-OH terminus. RNase P can process several other types of naturally occurring RNAs [61–63].

5.14.4   HEPATITIS DELTA VIRUS (HDV) RIBOZYME

The HDV Ribozyme is a class II ribozyme, and like the hairpin ribozyme, it is present in circular subviral RNAs for processing genomic units during rolling circle replication [64].

5.14.5   HAMMERHEAD RIBOZYME

The Hammerhead Ribozyme is a RNA molecule motif that catalyzes reversible cleavage and joining reactions at a specific site within an RNA molecule. It serves as a model system for research on the structure and properties of RNA, and it is used for targeted RNA cleavage experiments. They were reported in virioids, newt, *Arabidopsis*, crickets, and humans [64].

The hammerhead ribozyme carries out a very simple chemical reaction that results in the breakage of the substrate strand of RNA, specifically at C17, the cleavage-site nucleotide. Structurally the hammerhead ribozyme is composed of three base-paired helices, separated by short linkers of conserved sequences. These helices are called I, II and III. If the 5' and 3' ends of the sequence contribute to stem I, then it is a type I hammerhead ribozyme; to stem II, then it is a type II; and to stem III, then it is a type III hammerhead ribozyme [64].

5.14.6   VARKUD SATELLITE (VS) RIBOZYME

The Varkud Satellite (VS) Ribozyme is a large, naturally occurring nucleolytic ribozyme encoded by the Varkud satellite plasmid, found only in the mitochondria of certain *Neurospora* isolates. The enzyme has both cleavage and ligation activity and can perform both cleavage and ligation reactions efficiently in the absence of proteins [65].

5.14.7 HAIRPIN RIBOZYME

The Hairpin Ribozyme is a small section of RNA that can act as a ribozyme. It is found in RNA satellites of plant viruses. The hairpin ribozyme is an RNA motif that catalyzes RNA processing reactions essential for replication of the satellite RNA molecules in which it is embedded. Such reactions are important for processing the large multimeric RNA molecules.

A new riboswitch was found in a conserved region upstream of the *Bacillus subtilis* GlmS gene that regulates not through a conformational change but through ribozyme activity [66]. Binding of the metabolite glucosamine 6-phosphate (GlcN6P) to this riboswitch results in a site-specific mRNA cleavage reaction, thereby downregulating the expression of the GlcN6P synthetase GlmS ribozyme and further GlcN6P synthesis. The GlmS ribozyme promotes attack by a 2'-OH on its own 3'-phosphodiester bond, making it a class II ribozyme like the HammerHead, Hairpin, HDV, and VS ribozymes [66].

5.14.8 ARTIFICIAL RIBOZYMES

Artificial Ribozymes have been engineered that have catalyzed a spectrum of reactions like RNA ligation, RNA capping, RNA polymerization, cofactor synthesis, Dies-Alder reaction, urea synthesis, peptide bond formation, purine and pyrimidine synthesis, etc.

5.14.9 ALLOSTERIC RIBOZYMES

Allosteric Ribozymes are a modified type of catalytic RNA whose activity can be regulated by external factors. Control of their cleavage activity is achieved by binding an effector molecule to an allosteric binding site. This strategy can be exploited in situations in which a regulation of RNA-cleavage activity by an external factor is desirable, e.g., cleavage of mRNAs or viral RNAs in infected cells. Maxizymes are allosteric ribozymes whose activity is regulated by a specific sequence in the target mRNA and is active only when they form a dimer. The maxizyme is an allosterically controllable ribozyme with powerful biosensor capacity that appears to function even in mice [67].

5.14.10 DEOXYRIBOZYMES

Deoxyribozymes, also called DNA enzymes, DNAzymes, or catalytic DNA, are oligonucleotides that are capable of performing a specific targeted reaction. The most abundant class of deoxyribozymes are ribonucleases, which catalyze the cleavage of a ribophosphodiester bond through a transesterification reaction, forming a 2'3'-cyclic phosphate terminus and a 5'-hydroxyl terminus.

To date, no naturally occurring DNA-based enzyme has been discovered, and they have been shown capable of catalyzing metal-responsive phosphodiester bond cleavage and ligation, amide bond synthesis, polymerase activity, Diels–Alderase activity, porphyrin metalation, and others. They also can show RNA ligation and DNA phosphorylation [60–64, 68, 69].

Daniel and Sen [69] reported that UV1C deoxyribozyme could catalyze photochemical repair reaction of thymine dimers [68, 69].

5.14.11 POTENTIALS

- *Analytical sensors*: Lu et al. have developed deoxyribozyme sensors for metal ions and small organic molecules whose sensing ability is based on fluorescence or colorimetric signals [68–70]. Lan et al. reported that because of their high metal ion selectivity, one of the most important practical applications for DNAzymes is metal ion detection, resulting in highly sensitive and selective fluorescent, colorimetric, and electrochemical sensors for a wide range of metal ions such as Pb(2+) and $UO_2(2+)$. Such sensors can be employed in the testing of water for contamination [71].
- *Molecular robots*: Lund et al. [72] have developed autonomous DNA molecules called walkers or DNA spiders. When using appropriately designed DNA origami, the molecular spiders autonomously carry out sequences of actions such as "start," "follow," "turn," and "stop."

- *Lab work*: RNA-cleaving deoxyribozymes have been particularly useful as in vitro laboratory reagents for RNA cleavage [73]. Alternatively, chemically modified ribozymes may also be employed [74]. Such approaches using nucleic acid enzymes are a valuable counterpart to other mRNA-targeting strategies, most notably the application of small interfering RNA (siRNA) or antisense oligonucleotides.
- *RNA ligation*: As deoxyribozymes can show RNA ligation, this application can be used to join RNA molecules. The catalysis of the ligation of nucleic acid-based substrates is of importance for biochemical and biotechnological applications because this represents a valid alternative to splint ligations and to solid-phase synthesis.
- *Catalysis*: A few reactions catalyzed by catalytic nucleic acids have been summarized in the proceeding paragraphs.
- *DNA cleavage*: DNA-cleaving DNAzymes were discovered by Silverman et al. [68, 74].
- *Therapeutic applications*: By modifying the substrate recognizing sequences, catalytic nucleic acids can be specifically tailored for the suppression of particular genes. A large variety of gene products that are responsible for different pathological conditions have been targeted successfully using this strategy [72–77].
- The first clinical trial using a ribozyme targeted human immunodeficiency virus 1(HIV-1) had shown that an anti-HIV-1 gag ribozyme that was delivered intracellularly can interfere with both preintegration and postintegration events of the HIV replication cycle, by cleaving incoming viral RNA and transcribed mRNAs [78–81].

5.15 SUMMARY

Enzymes have been employed in the industries from creating lactose-free dairy products to fast-acting laundry detergents, to the textiles, foods, detergents, animals, biofuels and also find application in therapeutics and the healthcare industry, and in recombinant DNA technology. Nanobiosensors are the sensors made up of nanomaterials that have one of their dimensions between 1 and 100 nanometers. Due to the threat of bacterial resistance, the quest for new antimicrobial agents or drug targets prompted research in short peptides called antimicrobial peptides, which are divided into subgroups on the basis of their amino acid composition and structure. Peptide antibiotics are produced by bacterial, mammalian, insect, and plant organisms in defense against invasive microbial pathogens. Catalytic antibodies can enhance a couple of chemical and metabolic reactions in the body by binding a chemical group, resembling the transition state of a given reaction. An important point to notice is that ordinary antibodies do not chemically alter an antigen they bind to and neutralize. Ribozymes are nucleic acid molecules that catalyze important cellular reactions like intronic cleavages, cleavage of a specific phosphodiester bond, etc.

5.16   REVIEW QUESTIONS

1. How is the formation of the juice from the fruit pulps facilitated?
2. How can the enzymes be used to act as detergents? How are they better than chemical detergents?
3. What are biosensors and nanobiosensors?
4. Can you list a few enzymes for use clinically?
5. Are nucleases enzymes? Can you list a few important enzymes used in recombinant DNA technology?
6. What are AMPs?
7. How do the nucleic acids act as enzymes?

KEYWORDS

- acoustic wave biosensors
- biosensors
- enzymes
- ligases
- nanobiosensors
- nucleases
- polymerases
- peptides
- ribozyme

REFERENCES

1. Maarel, M., Veen, B. V. D., Uitdehaag, J. C. M., Leemhuis, H., & Dijkhuizen, L., (2002). Properties and applications of starch-converting enzymes of the –amylase family. *J. Biotechnol., 94*(2). DOI: 10.1016/S0168-1656(01)00407-2.
2. Kuraishi, C., Yamazaki, K., & Susa, Y., (2001). Transglutaminase: Its utilization in the food industry. *Food Rev. Int., 17*(2), 221–246.
3. Butt, S. M., Tahir-Nadeem, M., Ahmad, Z., & Sultan, M. T., (2008). Xylanases and their applications in baking industry. *Food Technol. Biotechnol., 46*(1), 22–31.
4. Malakar, R., Tiwari, A., & Malviya, S. N., (2010). Pullulanase: A potential enzyme for industrial application. *Int. J. Biomed. Res., 1*(2), 10–20.
5. Bhosale, S. H., Rao, M. B., & Deshpande, V. V., (1996). Molecular and industrial aspects of glucose isomerase. *Microbiol. Rev., 60*(2), 280–300.
6. Van Donkelaar, L. H. G., Mostert, J., Zisopoulos, F. K., Boom, R. M., & Van Der Goot, A. J., (2016). The use of enzymes for beer brewing: Thermodynamic comparison on resource use. *Energy, 115*(1), pp. 519–527.
7. Araujo, R., Casal, M., & Cavaco-Paulo, A., (2008). Application of enzymes for textiles fibers processing. *Biocatal. Biotechnol., 26*, 332–349.
8. Tzanov, T., Calafell, M., Guebitz, G. M., & Cavaco-Paulo, A., (2001). Bio-preparation of cotton fabrics. *Enzyme Microb. Technol., 29*, 357–362.
9. Adrio, J. L., & Demain, A. L., (2014). Microbial enzymes: Tools for biotechnological processes. *Biomolecules, 4*(1), 117–139. http://doi.org/10.3390/biom4010117.
10. Hasan, F., Shah, A. A., Javed, S., & Hameed, A., (2010). Enzymes used in detergents: Lipases. *Afr. J., Biotechnol., 9*(31), pp. 4836–4844.
11. Vulfson, E. N., (1994). *Industrial Applications of Lipases in Lipases*, Wooley, P., & Petersen, S. B., eds., Cambridge University Press, Cambridge, Great Britain, pp. 271.
12. Kumar, C. G., Tiwari, M. P., & Jany, K. D., (1999). Novel alkaline serine proteases from alkaliphilic *Bacillus* species: purification and characterization. *Proc. Biochem., 34*, 441–449.
13. Singhania, R. R., Patel, A. K., Thomas, L., Goswami, M., Giri, B. S., & Pandey, A., (2015). *Industrial Enzymes*. Elsevier Publications, pp. 473–497.
14. Adrio, J. L., & Demain, A. L., (2014). Microbial enzymes: Tools for biotechnological processes. *Biomolecules, 4*(1), 117–139.
15. Vanhanena, M., Tuomia, T., Tiikkainena, U., Tupaselaa, O., Tuomainenb, A., Luukkonena, R., & Nordmana, H., (2001). Sensitisation to enzymes in the animal feed industry. *Occup. Environ. Med., 58*, 119–123.
16. Li, S., Yang, X., Yang, S., Zhu, M., & Wang, X., (2012). Technology prospecting on enzymes: application, marketing and engineering. *Comput. Struct. Biotechnol. J., 2*(3), 2, e201209017. http://doi.org/10.5936/csbj.201209017.
17. De Souza, F. R., & Gutterres, M., (2012). Application of enzymes in leather processing: a comparison between chemical and coenzymatic processes. *Braz. J. Chem. Eng., 29*(3), 473–481.
18. Jaouadi, N. Z., (2015). A novel keratinase from *Bacillus tequilensis* strain Q7 with promising potential for the leather bating process. *Int. J., Biol. Macromolec., 79*, 952–964.

19. Thevenot, D. R., Toth, K., Durst, R. A., & Wilson, G. S., (2001). Electrochemical biosensors: recommended definitions and classification. *Biosens. Bioelectron., 16*(1–2), 121–131.

20. Villaverde, A., (2003). Allosteric enzymes as biosensors for molecular diagnosis. *FEBS Letters, 554*(1–2), 169–172.

21. Yoo, E. H., & Lee, S. Y., (2010). Glucose biosensors: An overview of use in clinical practice. *Sensors (Basel), 10*(5), 4558–4576.

22. Weibel, M. K., & Bright, H. J., (1971). The glucose oxidase mechanism. Interpretation of the pH dependence. *J. Biol. Chem., 246*, 2734–2744.

23. Guilbault, G. G., & Lubrano, G. J., (1973). An enzyme electrode for the amperometric determination of glucose. *Anal. Chim. Acta., 64*, 439–455.

24. Malik, P., Katyal, V., Malik, V., Asatkar, A., Inwati, G., & Mukherjee, T. K., (2013). Nanobiosensors: Concepts and variations. *ISRN Nanomaterials*, Article ID 327435, pp. 9.

25. Rai, M., Gade, A., Gaikwad, S., Marcato, P. D., & Durán, N., (2012). Biomedical applications of nanobiosensors: the state-of-the-art. *J. Braz. Chem. Soc., 23*(1), 14–24.

26. Shruthi, G. S., Amitha, C. V., & Mathew, B. B., (2014). Biosensors: A modern-day achievement. *Journal of Instrumentation Technology, 2*(1), 26–39.

27. Kulkarni, A. S., Joshi, D. C., & Tagalpallewar, G. P., (2014). Biosensors for food and dairy industry. *Asian J. Dairy & Food Res., 33*(4), 292–296.

28. Gooding, O., Voladri, R., Bautista, A., Hopkins, T., Huisman, G., Jenne, S., Ma, S., Mundorff, E. C., Savile, M. M., & Truesdell, S. J., (2010). Development of a practical biocatalytic process for (R)-2-methylpentanol. *Org. Process Res. Dev., 14*, 119–126.

29. Savile, C. K., Janey, J. M., Mundorff, E. C., Moore, J. C., Tam, S., Jarvis, W. R., Colbeck, J. C., Krebber, A., Fleits, F. J., Brands, J., et al., (2010). Biocatalytic asymmetric synthesis of chiral amines from ketones applied to sitagliptin manufacturer. *Science, 329*, 305–309.

30. Liang, J., Lalonde, J., Borup, B., Mitchell, V., Mundorff, E., Trinh, N., Kochreckar, D. A., Cherat, R. N., & Ganesh, P. G., (2010). Development of a biocatalytic process as an alternative to the (-)-DIP-Cl-mediated asymmetric reduction of a key intermediate of Montelukast. *Org. Process Res. Dev., 14*, 193–198.

31. Lods, L. M., Dres, C., Jhonson, C., Scholz, D. B., & Brooks, G. J., (2000). The future of enzymes on cosmetics. *Int. Jour. Cosmetic Sci., 22*, 8–94.

32. Clark, L., & Dekker, M., (1989). *Genetic Engineering Fundamentals.* An introduction to principles and applications. New York, pp. 290.

33. Russell, D. W., & Joseph, S., (2001). *Molecular Cloning: A Laboratory Manual*, vol. *1*. Third Ed. CSHL Press.

34. Dubos, R. J., (1939). Studies on a bactericidal agent extracted from a soil bacillus: I. Preparation of the agent. Its activity *in vitro. J. Exp. Med., 70*, 1–10.

35. Lai, R., Liu, H., Lee, W. H., & Zhang, Y., (2002). An anionic antimicrobial peptide from toad Bombina maxima. *Biochem. Biophys. Res. Comm., 295*, 796–799.

36. Stotz, H. U., Thomson, J. G., & Wang, Y., (2009). Plant defensins: Defense, development and application. *Plant Signaling & Behavior, 4*(11), 1010–1012.

37. Wang, G., (2014). Human antimicrobial peptides and proteins. *Pharmaceuticals, 7*(5), 545–594. http://doi.org/10.3390/ph7050545.

38. Castle, M., Nazarian, A., Yi, S. S., & Tempst, P., (1999). Lethal effects of apidaecin on *Escherichia coli* involve sequential molecular interactions with diverse targets. *J. Biol. Chem., 274*, 32555–32564.

39. Bahar, A. A., & Ren, D., (2013). Antimicrobial peptides. *Pharmaceuticals (Basel), 6*(12), 1543–1575.

40. Papagianni, M., (2003). Ribosomally synthesized peptides with antimicrobial properties: biosynthesis, structure, function, and applications. *Biotechnol. Adv., 21*(6), 465–499.

41. Gevers, W., Kleinkauf, H., & Lipmann, F., *(1968)*. The activation of amino acids for biosynthesis of gramicidin S. *Proc. Natl. Acad. Sci. USA, 60*(1), *269*–276.

42. Yukioka, M., & Winnick, T., (1966). Biosynthesis of malformin in washed cells of *Aspergillus niger. 119*(3). *Biochimica et Biophysica Acta (BBA) – Nucleic Acids and Protein Synthesis*, pp. 614–623.

43. Atkinson, D. J., Naysmith, B. J., Furkert, D. P., & Brimble, M. A., (2016). Beilstein enduracididine, a rare amino acid component of peptide antibiotics: Natural products and synthesis. *J. Org. Chem., 12*, 2325–2342.

44. Jenssen, H., (2008). QSAR modeling and computer-aided design of antimicrobial peptides. *J. Pept. Sci., 14*(1), 110–114.

45. Cherkasov, A., Hilpert, K., Jenssen, H., Fjell, C. D., Waldbrook, M., Mullaly, S. C., Volkmer, R., & Hancock, R. E., (2009). Use of artificial intelligence in the design of small peptide antibiotics effective against a broad spectrum of highly antibiotic-resistant superbugs. *ACS Chem. Biol., 4*(1), 65–74.

46. Blackburn, G. M., Kang, A. S., Kingsbury, G. A., & Burton, D. R., (1989). Catalytic antibodies. *Biochem. J., 262*, 381–390.

47. Paul, S., Volle, D. J., Beach, C. M., Johnson, D. R., Powell, M. J., & Massey, R. J., (1989). Catalytic hydrolysis of vasoactive intestinal peptide by human autoantibody. *Science, 244*, 1158.

48. Paul, S., Karle, S., Planque, S., Taguchi, H., Salas, M., Nishiyama, Y., Handy, B., Hunter, R., Edmundson, A., & Hanson, C., (2004). Naturally occurring proteolytic antibodies: Selective immunoglobulin M-catalyzed hydrolysis of HIV gp120. *J. Biol. Chem., 279*(38), 39611–39619.

49. Benkovic, S. J., (1992). Catalytic antibodies. *Annu. Rev. Biochem., 61*, 29–54.

50. Padiolleau-Lefevre, S., Naya, R. B., Shahsavarian, M. A., Friboulet, A., & Avalle, B., (2014). Catalytic antibodies and their applications in biotechnology: state of the art. *Biotechnol. Lett., 36*, 1369–1379.

51. Landry, D. W., Zhao, K., Yang, G. X. Q., Glickman, M., & Georgiadis, T. M., (1993). *Science, 259*, 1899–1901.

52. Suzuki, H., (1994). Recent advances in abzyme studies. *J. Biochem., 115*(4), 623–628.

53. Paul, S., Volle, D. J., Beach, C. M., Johnson, D. R., Powell, M. J., & Massey, R. J., (1989). *Science, 244*, 1158–1162.

54. Dickerson, T. J., Yamamoto, N., & Janda, K. D., (2004). Antibody-catalyzed oxidative degradation of nicotine using riboflavin. *Bioorg. Med. Chem., 12*(18), 4981–4987.

55. Shabat, D., Lode, H. N., Pertl, U., Reisfeld, R. A., Rader, C., Lerner, R. A., & Barbas, C. F., (2001). *In vivo* activity in a catalytic antibody-prodrug system: Antibody cata-

lyzed etoposide prodrug activation for selective chemotherapy. *Proc. Natl. Acad. Sci. USA, 98*(13), 7528–7533.

56. Wentworth, P., Datta, A., Blakey, D., Boyle, T., Partridge, L. J., & Blackburn, G. M., (1996). Toward antibody-directed "abzyme" prodrug therapy, ADAPT: carbamate prodrug activation by a catalytic antibody and its in vitro application to human tumor cell killing. *Proc. Natl. Acad. Sci. USA, 93*(2), 799–803.

57. Shabat, D., Lode, H. N., Pertl, U., Reisfeld, R. A., Rader, C., Lerner, R. A., & Barbas, C. F., (2001). *In vivo* activity in a catalytic antibody-prodrug system: Antibody catalyzed etoposide prodrug activation for selective chemotherapy. *Proc. Natl. Acad. Sci. USA, 98*(13), 7528–7533. http://doi.org/10.1073/pnas.131187998.

58. Zorrilla, E. P., (2006). Vaccination against weight gain. *Proc. Natl. Acad. Sci., 103*(35), 13226–13231.

59. Paul, S., Planque, S., Zhou, Y. X., Taguchi, H., Bhatia, G., Karle, S., Hanson, C., & Nishiyama, Y., (2003). *J. Biol. Chem., 278*(22), 20429–20435.

60. Scott, K., & Silverman, S. K., (2008). Nucleic acid enzymes (*Ribozymes and Deoxyribozymes*): In vitro Selection and Application, University of Illinois at Urbana-Champaign, Urbana, Illinois, Begley, T. P., (ed.), *Wiley Encyclopedia of Chemical Biology*, John Wiley & Sons, Hoboken (NJ, USA).

61. Ward, W. L., Plakos, K., & DeRose, V. J., (2014). Nucleic acid catalysis: Metals, nucleobases, and other cofactors. *Chem. Rev., 114*, 4318–4342.

62. Lambowitz, A. M., Caprara, M. G., Zimmerly, S., & Perlman, P. S., (1999). *The RNA World, 2nd* ed., Cold Spring Harbor Laboratory Press, 0-87969-561-7/99.

63. Lan, T., & Lu, Y., (2012). Metal Ion-Dependent DNAzymes and Their Applications as Biosensors. In: *Metal Ions in Life Sciences*, Springer: New York, *10*(167), Chapter 8, pp. 217–248.

64. Doherty, E. A., & Doudna, J. A., (2001). Ribozyme structures and mechanisms. *Annu. Rev. Biophys. Biomol. Struct., 30*, 457–475.

65. Lilley, D. M. J., (2004). The Varkud satellite ribozyme. *RNA, 10*(2), 151–158. http://doi.org/10.1261/rna.5217104.

66. Caron, M. P., Bastet, L., Lussier, A., Simoneau-Roy, M., Massé, E., & Lafontaine, D. A., (2012). Dual-acting riboswitch control of translation initiation and mRNA decay. *Proceedings of the National Academy of Sciences of the United States of America, 109*(50), 3444–3453. http://doi.org/10.1073/pnas.1214024109.

67. Hamada, M., Kuwabara, T., Warashina, M., Nakayama, A., & Taira, K., (1999). Specificity of novel allosterically *trans*- and *cis*-activated connected maxizymes that are designed to suppress *BCR-ABL* expression. *FEBS Letters, 461*(1–2), 12, 77–85.

68. Silverman, S. K., (2010). DNA as a versatile chemical component for catalysis, encoding, and stereocontrol. *Angew. Chem., Int. Ed., 49*, 7180.

69. Chinnapen, D. J. F., & Sen, D., (2004). A deoxyribozyme that harnesses light to repair thymine dimers in DNA." *Proc. Natl. Acad. Sci. USA, 101*(1), 65–69.

70. Liu, J., & Lu, Y., (2003). A colorimetric lead biosensor using DNAzyme directed assembly of gold nanoparticles. *J. Am. Chem. Soc., 125*, 6642–6643.

71. Liu, J., & Lu, Y., (2004). Adenosine-dependent assembly of aptazyme functionalized gold nanoparticles and its application as a colorimetric biosensor. *Anal. Chem., 76*, 1627–1632.

72. Liu, J., & Lu, Y., (2006). Fast colorimetric sensing of adenosine and cocaine based on a general sensor design involving aptamers and nanoparticles. *Angew. Chem. Int. Ed.*, *45*, 90–94.

73. Lan, T., & Lu, Y., (2012). Metal ion-dependent DNAzymes and their applications as biosensors. In: *Metal Ions in Life Sciences*, Springer: New York, vol. *10*, Chapter 8, pp. 217–248.

74. Silverman, S. K., (2005). In vitro selection, characterization, and application of deoxyribozymes that cleave RNA. *Nucleic Acids Res.*, *33*, 6151–6163.

75. Tafech, A., Bassett, T., Sparanese, D., & Lee, C. H., (2006). Destroying RNA as a therapeutic approach. *Curr. Med. Chem.*, *13*, 863–881.

76. Chandra, M., Sachdeva, A., & Silverman, S. K., (2009). DNA-catalyzed sequence-specific hydrolysis of DNA. *Nat. Chem. Biol.*, *5*, 718–720.

77. Marcel Hollenstein. *DNA Catalysis*: The chemical repertoire of DNAzymes.

78. Trang, P., Kilani, A., Lee, J., Hsu, A., Liou, K., Kim, J., et al., (2002). RNase P ribozymes for the studies and treatment of human cytomegalovirus infections. *J. Clin. Virol.*, *25*, 63–74.

79. Konopka, K., Rossi, J. J., Swiderski, P., Slepushkin, V. A., & Duzqunes, N., (1998). Delivery of an anti-HIV-1 ribozyme into HIV-infected cells via cationic liposomes. *Biochim. Biophys. Acta.*, *1372*, 55–68.

80. Abera, G., Berhanu, G., & Tekewe, A., (2012). Ribozymes: Nucleic acid enzymes with potential pharmaceutical applications – A review. *Pharmacophore*, *3*(3), 164–178.

CHAPTER 6

WHITE BIOTECHNOLOGY

PRERNA PANDEY and ANJALI PRIYADSRSHINI

CONTENTS

6.1   INTRODUCTION

Since a long time, humans relied on nature to provide them with various products to make life more comfortable. But with the passage of time, the use of plastics and other synthetic materials predominated over natural resources. With growing concerns about the environment, people are looking at natural or biodegradable options. Biotechnology has found its entry into medicine (red) and agriculture (green), and now, a new wave of modern biotechnology is gaining momentum – "white biotechnology" is the application of nature's toolset to industrial production.

The term was defined by Karl Erich Jaeger as biotechnological production of compounds with the help of enzymes or microbes. An entire branch of biotechnology, known as "white biotechnology," is devoted to this. It employs living cells—from yeast, molds, bacteria, and plants—and enzymes to synthesize products that are easily degradable, require less energy, and create less waste during their production.

The principle of "white biotechnology" is not new. Bacterial enzymes have been used widely in food manufacturing and as active ingredients in washing powders to reduce the number of artificial surfactants. Transgenic *Escherichia coli* is used to produce human insulin in large-scale fermentation tanks for use in therapeutic purposes.

Biocatalysts (enzymes and microorganisms) are the key tools of white biotechnology, which are considered to be one of the key technological drivers for the growing bio-economy. The advantages offered are many–the dependence on natural resources can reduce dependence on depleting and nonrenewable fossil fuels. The substrates and waste are biologically degradable, which aids to decrease their environmental impact.

Substrate specific synthesis has been found to employ mild conditions. The usage of alternative and renewable sources helps in reducing the carbon foot print. Use of enzymes for bioprocessing has various benefits in contrast to traditional techniques and methods being used in industry. Because enzymes are like catalysts in a reaction, they always remain intact. They also possess optimum temperature and optimum pH and hence are easy to control in any given reaction. The biggest advantage of using enzymes is that it is absolutely biodegradable and shuns the use of otherwise needed chemicals and toxic substances to perform textile processes. The use of enzymes aid in reducing the consumption of excess water and energy and also does not cause any pollution.

6.2 CASE STUDIES

6.2.1 BIO-BASED FIBERS

Dupont (Wilmington, DE, USA), the company that invented nylon, has for many years been developing a polymer based on 1,3-propanediol (PDO), with new levels of performance, resilience, and softness. By using glucose as the basis for Bio-PDO™, a bio-based monomer, DuPont created a renewably sourced ingredient for bio-based fibers, like DuPont™ Sorona (Figure 6.1), which is used in everyday products such as carpet and apparel. With Bio-PDO™, DuPont has found another way to reduce dependence on fossil fuels. The production of Bio-PDO™ consumes up to 40% less energy and reduces greenhouse gas emissions by more than 40% versus petroleum-based PDO. Using Bio-PDO™ as a key ingredient, the production of Sorona reduces greenhouse gas emissions by 63% over petroleum-based nylon [6].

Apparel manufacturers appreciate that Sorona combines the best of both nylon and polyester in one fiber, delivering extraordinary softness, exceptional comfort stretch, brilliant color, and easy care. All of these

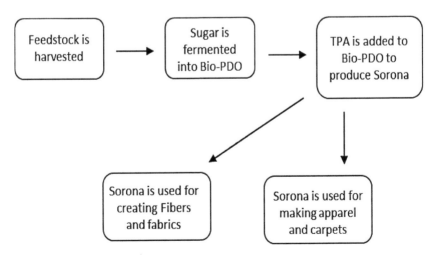

FIGURE 6.1 Sorona is made partly with plant-based ingredients that are annually renewable.

attributes enhance the look, feel, and quality of active wear, swimwear, intimates, denim, and other apparel [1].

6.2.2   BIOPOLYMER

Cargill Dow (Minnetonka, MN, USA) has gone a step further. The company has developed an innovative biopolymer, NatureWorks™, which can be used to manufacture items such as clothing, packaging, and office furnishings. The polymer is derived from lactic acid, which is obtained from the fermentation of corn sugar. The commercial quality polymer is made from the carbon found in simple plant sugars such as corn starch to create a proprietary *polylactic* acid polymer (PLA), which is marketed under the brand name Ingeo. Headquartered in Minnetonka, Minnesota, NatureWorks is jointly owned by Cargill and PTT Global Chemical (Figure 6.2).

Ingeo biopolymer: This polylactide is derived from dextrose sugar derived from corn. The dextrose is converted to lactic acid and then lactide

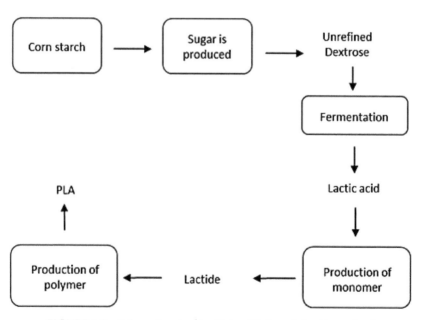

FIGURE 6.2   Schematics showing NatureWorks polylactide production.

rings. This finds application in food containers, food packaging, electronics, films, and apparels [2].

6.3   ENZYME-BASED WHITE BIOTECHNOLOGY

Henkel employs white biotechnology in the use of enzymes for laundry detergents and automatic dish washing detergents. These enzymes show significantly better environmental performance, even during their production, than conventionally produced enzymes do. Besides enzymes, Henkel's laundry and home care products contain further ingredients based on white biotechnology, e.g., biosurfactants and citric acid [3]. Various substances/chemicals are produced employing white biotechnology:

6.3.1   VITAMINS

Traditionally, Vitamin B2 is produced using a complex eight-step chemical process. BASF's new biotech process reduces production to a one-step process. This single step is a fermentation whereby the raw material is fed to a microorganism – in this case mold – that transforms it into the finished product. Vitamin B2 is recovered as yellow crystals directly from the fermentation. Using the fungus *Ashbya gossypii* as a biocatalyst, BASF achieved an overall reduction in cost and environmental impact of 40% [4].

6.3.2   MEDICINE

Another good example is DSM's route to the antibiotic Cephalexin, practiced on an industrial scale for several years. The complex, traditional chemical process involved many steps. Metabolic pathway engineering helped to establish a mild bio-transformation route that has reduced the process steps substantially. The new route is based on a fermented intermediate linked enzymatically with a side chain to the final end product. The biotechnological process uses less energy and less input chemicals, is water-based, and generates less waste in the process [4].

6.3.3   BIOFUELS

Biofuels are liquid or gaseous fuels that are produced from biodegradable fractions of products, remains from agricultural production and forestry, and biodegradable fractions of industrial and municipal wastes.

Ethanol is currently used in the fuel industry as an additive for petrol. It is a high-octane fuel and has replaced lead as an octane enhancer in petrol. Blending ethanol with petrol oxygenates the fuel mixture so that it burns completely and reduces harmful emissions. The most common blend is 90% petrol and 10% ethanol [4]. Bioethanol has a number of benefits when compared to conventional fuels. Firstly, it is produced from a renewable resource (such as crops). Blending bioethanol with petrol compensates for the diminishing oil supplies across the globe thereby ensuring higher fuel security and avoiding foreign reliance for fuel supply between countries. The rural community will also benefit from the increased demand to grow the necessary crops required for producing bioethanol.

An efficient bioethanol production by *Bacillus cereus* GBPS9 using sugarcane bagasse and cassava peels as feedstocks was demonstrated by Ezebuiro, et al. (Figure 6.3) [5]. The authors quote the rationale that

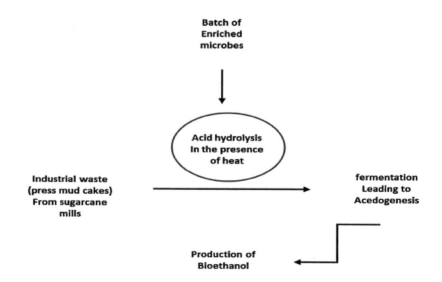

FIGURE 6.3   A simplified schematic of bioethanol production from sugarcane bagasse [5].

cassava can be grown in arid, marginal soil where other crops such as corn, sugarcane, and sugar beet fail.

Another key benefit of bioethanol is the ease of integrating it with the existing road transport fuel system – bioethanol can be easily blended with conventional fuels (up to 15%) without any need for engine modifications.

6.3.4   BIOPLASTICS

Polyhydroxyalkanoates (PHAs) are a family of naturally synthesized polymer in a wide variety of bacteria as an energy and/or carbon storage material. Polyhydroxyalkanoates or PHAs are linear polyesters produced by fermentation by bacteria. These polyesters are deposited intracellularly as energy deposits. The monomers may be 3/4/5-hydroxy alkanoic acids or substituted versions. A variety of homo or copolymers can be constructed from these various monomers offering a wide possibility in the polymer production.

Most plastics are derived from finite sources of petroleum or natural gas. Traditional petro-based compounds are also challenging to deal with at end of use. They are not biodegradable and plastic products often involve complex and almost inseparable mixes of materials. Even when recycling is possible, the process often involves a downgrading of the quality. The properties of PHA are similar to those of polypropylene, which is usually found in packaging, carpets, stationery, and several other everyday products. It was found that organic residual streams, including sewage, contain all the necessary ingredients for biopolymers to be created. Since 2007, in the laboratories of Anoxkaldnes, the creation of polymers by bacteria in wastewater has been advancing to practical reality.

The first PHA that was used was for a shampoo bottle: a PHB (poly-hydroxybutyrate)/PHV (polyhydroxy-valerate) copolymer developed by ICI- now under Metabolix.

PHB is produced by microorganisms (such as *Ralstonia eutrophus*, *Bacillus megaterium, or Methylobacterium rhodesianum*).

PHA production can be achieved in open, mixed microbial cultures and thereby coupled to wastewater and solid residual treatment. In 2014, Morgan-Sagastume et al. reported the evaluation of feasibility of PHA production in processes for sludge treatment, volatile fatty acid (VFA) production, and municipal wastewater treatment [6]. Laboratory- and pilot-scale studies demonstrated the feasibility of municipal wastewater and solid waste treatment for the production of PHA-rich biomass. Volatile fatty acids (VFAs) obtained from waste-activated sludge fermentation were found to be a suitable feedstock for PHA production.

PHAs possess material properties similar to various petrochemical-based synthetic thermoplastics and elastomers currently in use. They are also completely degradable to water and carbon dioxide (and methane under anaerobic condition) by microorganisms in various environments (Figure 6.4).

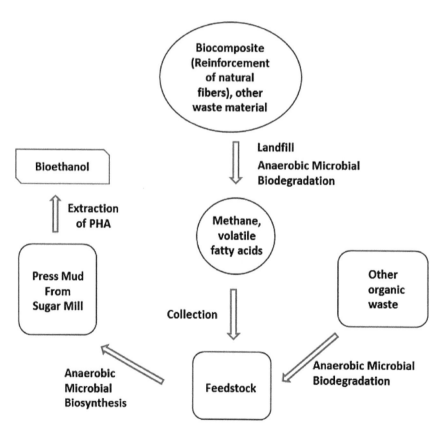

FIGURE 6.4 A schematic showing PHA production from volatile fatty acids (VFAs) obtained from waste activated sludge fermentation.

6.3.4.1 Applications of Bioplastics

Bioplastics can be used as packaging films (for food packages), bags, containers, and paper coatings. For food packaging applications, such as those used for produce and meats, PHAs have moisture vapor barrier properties [7] comparable to existing food-packaging materials such as polyethylene terephthalate and polypropylene. PHA materials also have a high surface energy, thus making the material receptive to printing inks and dyes [7].

In addition to its potential as a plastic material, PHA is useful source of stereoregular compounds that can serve as chiral precursors for the chemical synthesis of optically active compounds, particularly in synthesis of some drugs or insect pheromones [6, 7].

PHAs can be easily depolymerized to a rich source of optically pure bifunctional hydroxy acids. PHB, for example, can be readily hydrolyzed to R-3-hydroxybutyric acid and is used in the synthesis of Merck's antiglaucoma drug Truspot. Along with R-1,3-butanediol, it is also used to synthesize b-lactams.

Lactide was successfully copolymerized with commercial PHA polymer (Nodax™). Nodax is a family of bacterially produced PHA copolymers comprising 3-hydroxybutyrate (3HB) and other 3-hydroxyalkanoate (3HA) units with side groups greater than or equal to three carbon units for potential applications as films.

Due to its biocompatibility, the use of PHA bioplastics in healthcare delivery is a topic of research. The use of bioplastic products on live subjects has resulted in a much lower risk of rejection, inflammation, excessive scar tissue, and future physical damage than that caused by products made of metal or artificial plastics. They also show promise in cardiovascular field for use as stents and in tissue regeneration [6, 7].

6.4 OTHER CHEMICALS OF INTEREST

6.4.1 1,3-PROPANEDIOL

It has a wide range of applications such as composites, adhesives, laminates, monomers of polymers, solvents, and anti-freezing agent.

1,3-Propanediol (1,3-PD) is known as a trimethylene glycol, 1.3-dihydroxypropane, and propane-1,3-diol [8, 9]. The molecular formula of the compound is $C_3H_8O_2$ with applications as a monomer for the production of polyesters, polyethers, and polyurethanes. It is a raw material for the production of biodegradable plastics, films, solvents, adhesives, detergents, cosmetics, and medicines. Though 1,3-PD is produced in nature by bioconversion, the advent of biotechnology has helped in its production from glucose.

In the past, 1,3-PD was produced only chemically by two methods: the hydration of acrolein or the hydroformylation of ethylene, but the chemical synthesis had drawbacks of employing high pressure and temperature.

The earliest known 1,3-PD fermentation pathway was mentioned by August Freund [8] with glycerol and a mixed culture of microbes. Several bacterial genera are known for their production of PDO: *Klebsiella*, *Enterobacter*, *Citrobacter*, *Clostridium*, and *Lactobacillus*.

Glycerol is a renewable resource found as a by-product of ethanolic fermentation of glucose. Glycerol can be converted to 1,3-PD by many microorganisms such as *Klebsiella pneumoniae*, *Bacillus welchi*, *Lactobacillus* spp., *Enterobacter* spp., *Citrobacter* spp., and *Clostridia* spp. [9, 10].

The production of 1,3-PD from glycerol is performed under anaerobic conditions using glycerol as the sole carbon source in the absence of other exogenous reducing equivalent acceptors. However, strains like *K. pneumoniae*, *K. oxytoca*, *K. aerogenes*, *Enterobacter agglomerans*, *E. aerogenes*, *Citrobacter freundii*, *Lactobacillus reuteri*, *L. buchnerii*, *L. collinoides*, *Pelobacter carbinolicus*, and *Rautella planticola* have been found to produce 1,3-PD in microaerobic fermentation [11, 12].

The use of 1,3-PD in thermourethane has improved its properties. Glycerol is a byproduct of esterification of vegetable oils. Interfacial engineering and biotechnology (IGB) has developed microbes that can convert glycerol to 1,3-PD.

DuPont has a product, Susterra 1,3-propanediol, which finds potential in the manufacturing of polyester-based resins and as a cross-linker in urethane chemistries.

In addition, Susterra 1,3-propanediol can be used in applications such as pharmaceutical intermediates, deicing, and heat transfer fluids and as a base for engine coolants.

1,3-PD is also used in the production of polypropylene terephthalate (PPT or PTT or 3GT) that finds applications in the manufacture of carpets, clothing, textiles, engineering, thermoplastics, and apparel [10–13].

6.4.2 LACTIC ACID

IGB is used to produce lactic acid from whey. Whey is considered as a waste product once its protein products are removed. It mainly contains lactose that has poor sweetening properties. The employment of lactic acid bacteria can convert lactose to lactic acid that finds application as an acidulant and preservative [13].

3-Hydroxyproprionic acid is a key intermediate for several commercially important chemicals including acrylic acid, acrylamide, and 1,3-PD. Codexis and Cargill developed a microbial process that utilizes low-cost and clean agricultural feedstocks (corn sugar) to produce 3-hydroxyproprionic acid [13].

6.4.3 SUCCINIC ACID ($C_4H_6O_4$)

Also known as amber acid or butanedioic acid, is a four-carbon dicarboxylic acid that can be used as a precursor of many industrially important chemicals in food, chemical, and pharmaceutical industries. Succinic acid can be produced by different kinds of anaerobic and facultative bacteria as a fermentation end-product [4, 13]. However, only a few species can produce it as the major end-product with high yield, such as *Anaerobiospirullum succiniciproducens, Actinobacillus succinogenes, and Mannheimia succiniciproducens*. As *Escherichia coli* produces succinic acid as a minor fermentation product under anaerobic conditions, several metabolic engineering strategies have been employed for the enhanced production of succinic acid by *E. coli* with some good results and production yields [13–15].

In 2016, Sadhukhan et al. [14] reported the production of succinic acid from glycerol and obtained optimum yield, thus showing that even waste products can be used by white biotechnology to produce desirable products.

Dicarboxylic acids find use in novel polyamides and other polymers. Microorganisms produce these acids by a pathway called w-oxidation.

6.5   FLAVOR

The majority of industrial flavors is provided by higher plants from many different species. But, agriculture is dependent on several variables like climatic fluctuations and human interference like the usage of various agrochemicals that lead to contamination of products. Higher fungi show the largest potential for flavor formation; all kinds of volatiles including methyl ketones, lactones, phenols, phenylpropanoids, terpenes and terpenoids, and more have been found in fungi [15].

With a special focus on wine, glycosidases were described to liberate aroma active aglyca from their glycosidic precursors. Glycosidases from *Aspergilli* showed less inhibition by glucose, fructose, ethanol and low pH, and their application resulted in improved wine flavor, as confirmed by sensory tests. The same approach worked with vanilla beans and birch bark (for the release of raspberry ketone) [15].

Among the oxidoreductases used for flavor generation are alcohol dehydrogenases, peroxidases including chloroperoxidase, lipoxygenases, amine oxidases and vanillyl alcohol oxidase. Basidiomycetes have been found to be a rich source of novel oxidoreductases [15].

6.6   CHEMICAL REACTIONS

Ugi reaction is one of the more well-known multicomponent reactions. The classic Ugi reaction involves condensation of a primary amine, a carbonyl compound, carboxylic acid, and isocyanide. An important objective of white biotechnology has been to be able to operate chemical processes starting with a simple set of compounds obtained from renewable resources. Peptide bond formation starting with an amine, aldehyde and isocyanide (rather than two amino acids) has been documented.

Liu et al. [16] reported *asymmetric aldol reactions* between isatin derivatives with cyclic ketones catalyzed by nuclease p1 from *Penicillium citrinum* [16]. This is a green signal in the synthesis of pharmaceutically active compounds. The same enzyme was used earlier by Li et al. [17] for catalyzing asymmetric aldol reactions between aromatic aldehydes and cyclic ketones [17].

Zhou et al. [18] used α-amylase from *Bacillus subtilis* for catalyzing the *oxa-Michael/aldol condensation* for the synthesis of substituted chromene derivatives [18].

In the same year, Gao et al. [19] used *glucoamylase* from *Aspergillus niger* to *catalyze Henry's reaction*, which they described as efficient and scalable [19]. *Polyketide synthetase* was employed for preparative scale synthesis of 15 new aureothin (a shikimate-polyketide with antimicrobial and antitumor activities) analogs by Werneburg et al. [20]. It was reported to have lesser cytotoxicity and higher activity. This result involved the use of genetic engineering techniques to achieve the result.

Few biotransformations used in industries are listed in Table 6.1 [21].

6.7   THE FOOD INDUSTRY

The biotech products that we use every day at home have many industrial applications, of which some utilize enzymes to produce or make improvements in the quality of different foods. Enzymes have been applied in food processing to provide stable and good quality products, with more efficient production. They also reduce energy, water, and raw materials consumption and generate environmentally safe waste. The following enzymes have been used in the food industry:

6.7.1   LIPASE

It helps in the production of fats with improved qualities. It enhances flavor development and shortens the time for cheese ripening. Lipase is involved in the production of enzyme-modified cheese/butter from cheese curd or butterfat.

6.7.2   ALPHA-AMYLASE

Catalyzes starch to dextrins conversion in producing corn syrup. It also solubilizes carbohydrates found in barley and other cereals used in brewing.

TABLE 6.1   A Summary of Enzymes That are Used in Different Industries

Substrate: type of biotransformations	Enzyme: source
Diketones: Regio- and stereoselective reductions	Alcohol Dehydrogenase from *Lactobacillus, Rhodococcus, Thermoanaerobacter* etc.
Phenylacetaldoxime: Reduction reaction	Alcohol Dehydrogenase from *Rhodococcus, Ralstonia sp., Thermoanaerobacter* etc.
Salicylaldehyde, ethyl acetoacetate: Domino Knoevenagel/intramolecular transesterification	Alkaline protease from *Bacillus licheniformis* (BLAP)
Guaifenesin, vinyl acetate: Transesterification reaction	Acylase: Penicillin G
Aromatic aldehyde, acetyl acetone/Ethyl acetoacetate: Knoevenagel reaction	Papain from *Carica papaya* latex
Aldehyde, amine, isocyanide: Ugi reaction for peptide synthesis (MCR)*	Novozyme 435 (commercially available, mmobilized *Candida antarctica* lipase B)
Salicylaldehyde, methyl vinyl ketone: Domino oxa-Michael/aldol condensations	α-amylase from *Bacillus subtilis*
4-cyanobenzaldehyde, nitromethane: Henry reaction	Glucoamylase from *Aspergillus niger* (AnGA)
L-threonine and L-*allo*-threonine (for retro aldol reaction); D- and L-alanine (for transamination): Retro aldol and transamination reactions	Alanine racemase from *Tolypocladium inflatum*
Substituted benzalacetones and 1,3-cyclic diketones: Michael addition-cyclization cascade reaction	BioH esterase from *E.coli*
p-nitrobenzaldehyde and methyl vinyl ketone: Baylis-Hillman reaction	
Salicylaldehyde, acetophenone, methanol: Synthesis of substituted 2H-chromemes (MCR)*	Lipase from Porcine pancreatic
Urea, ethyl acetoacetate, vinyl acetate: Biginelli reaction (MCR)*	Trypsin from porcine pancreas
Aromatic aldehyde, benzyl amine, mercaptoacetic acid: Synthesis of 4-thiazolidinones	
Benzylamine, isobutyraldehyde, thioacetic acid, methyl 3-(dimethylamino)-2-isocyanoacrylate: Synthesis of 2,4-disubstituted thiazoles (MCR)*	Lipase from Porcine pancreatic (PPL)
4-chlorobenzaldehyde, cyclohexenone, 4-anisidine: Aza-Diels-Alder reaction	Lysozyme from Hen egg white (HEWL)

6.7.3   LACTASE

Catalyzes breakdown of lactose in whey products for manufacturing poly-lactide. It also helps in digestion of dairy products in individuals lacking lactase.

6.7.4   PAPAIN

It is popularly used as a meat tenderizer. Papain is also used in brewing to prevent chill-haze formation by digesting proteins that otherwise react with tannins to form insoluble colloids.

6.7.5   GLUCOAMYLASE

Conversion of dextrins to glucose in the production of corn syrup. Conversion of residual dextrins to fermentable sugar in brewing for the production of "light" beer.

6.7.6   BETA-GLUCANASE

It is used in breweries for breakdown of glucans in malt and other materials to aid in filtration after mashing.

6.7.7   ACETOLACTATE DECARBOXYLASE

This enzyme catalyzes conversion of acetolactate to acetoin, thus, helps in reducing maturation time in wine making.

6.7.8   CHYMOSIN

It is used in cheese making by curdling of milk through breaking down kappa-caseins.

6.7.9 MICROBIAL PROTEASES

It is used in processing of raw plant and animal protein and in production of fish meals, meat extracts, texturized proteins, and meat extenders.

6.7.10 PECTINASE

It is used in treatment of fruit pulp to facilitate juice extraction and for clarification and filtration of fruit juice.

6.7.11 GLUCOSE OXIDASE

It helps in conversion of glucose to gluconic acid to prevent Maillard reaction caused by high heat used in dehydration in products.

6.7.12 CELLULASE

It catalyzes conversion of cellulose waste to fermentable feedstock for ethanol or single-cell protein production and in cell wall degradation of grains to allow better extraction of cell contents and release of nutrients.

6.8 THE DAIRY INDUSTRY

6.8.1 CHYMOSIN

Milk contains proteins, specifically caseins, that maintain its liquid form. Proteases are enzymes that are added to milk during cheese production to hydrolyze caseins, specifically kappa casein, which prevents coagulation by stabilizing micelle formation. The general terms- rennet and rennin- are used for any enzyme used to coagulate milk. The most common enzyme isolated from rennet is chymosin.

6.8.2 PROTEASES

Cow milk also contains whey proteins such as lactalbumin and lactoglobulin. The denaturing of these whey proteins, using proteases, results in a creamier yogurt product. Cheese production also requires degradation of whey proteins.

During the production of soft cheeses, whey is separated from the milk after curdling and may be sold as a nutrient supplement for weight loss, bodybuilding, and lowering blood pressure. There have even been reports of use of dietary whey for cancer therapies and in inducing insulin production for people with type 2 diabetes. Proteases are used to produce hydrolyzed whey protein, which is whey protein broken down into shorter polypeptides. Hydrolyzed whey is also used to prepare supplements for medical uses and infant formulas because it is less likely to cause allergic reactions.

6.8.3 LACTASE

Lactase is a glycoside hydrolase enzyme that catalyzes breakdown of lactose into galactose and glucose. Without sufficient production of lactase enzyme in the small intestine, humans become lactose intolerant, and consuming milk and milk products causes discomfort in the digestive tract.

Lactase is used commercially to prepare lactose-free products and in the preparation of ice cream to make a sweeter and creamier product. Lactase is usually prepared from *Kluyveromyces* sp. of yeast and *Aspergillus* sp.

6.8.4 CATALASE

Hydrogen peroxide is a potent oxidizer and causes cytotoxicity. Catalase is used instead of pasteurization, when making certain cheeses such as Swiss, in order to preserve natural milk enzymes that are beneficial to the flavor development of cheese. These enzymes are destroyed by the high heat during pasteurization. However, the hydrogen peroxide residues in the milk must be removed because it inhibits the bacterial cultures that are required for the actual cheese production; hence, all traces of it must be removed. Catalase enzymes are typically obtained from bovine livers or microbial sources and are added to convert hydrogen peroxide to oxygen and water.

6.8.5   LIPASES

The milk fats are broken down, and characteristic flavors are imparted to cheese with the use of lipases. Stronger flavored cheeses, for example, Romano, Italian cheese, are prepared using lipases. The free fatty acids produced impart flavor when milk fats are hydrolyzed. Animal lipases are obtained from calf and lamb, and microbial lipase is obtained by the fungus *Mucor meihei* by fermentation. Although microbial lipases are available for cheese-making, they are less substrate specific, while the animal enzymes are more partial to short and medium-length fats. Hydrolysis of the shorter fats is preferred because it results in the desirable taste of many cheeses. Hydrolysis of the longer chain fatty acids can result in either loss of flavor or soapiness.

6.9   THE BREWERY INDUSTRY

Alcoholic beverages have been a part of human life for thousands of years. These beverages, for example, wine and beer, are produced through fermentation of sugars by yeast. While the mature grapes contain the sugars needed for the fermentation, barley contains starch that needs to be broken down to sugars for fermentation to make alcohol. Thus, a malting step is used in which those enzymes are produced that are needed for the degradation of starch into fermentable sugars.

The standard mashing for beer consists of several temperature steps, involving different malt enzyme activities. The optimal temperature for β-glucanases, the cell wall-degrading enzymes, is 45°C; for proteases, 52°C; for β-amylase, 63°C; and for α-amylase, 72°C. The last step in the mashing is inactivation of the enzymes at 78°C [22].

During mashing, the starch is degraded to dextrin and fermentable sugars. α-amylase hydrolyzes the α-1,4 linkages at random and liquefy the gelatinized starch. β-amylases are exo-enzymes that attack the liquefied starch chains, resulting in successive removal of maltose units from the nonreducing end. Amyloglucosidases increase the glucose content and at the same time, pentosanase and xylanases hydrolyze pentosans of malt, barley, wheat to improve extraction and beer filtration.

After mashing, yeast is added for the fermentation step. Fungal α-amylase increases maltose and glucose content; β-glucanase hydrolyzes glucans, reduces viscosity, and aids filtration; α-acetolactate- decarboxylase converts α-acetolactate to acetoin directly, resulting in the decrease in fermentation time by avoiding formation of diacetyl.

After fermentation, an enzyme solution for diacetyl control ensures better vessel utilization, saves energy, and produces a high-quality beer after a reduced maturation time.

Diacetyl gives beer an off-flavor like buttermilk; for maturing a beer the diacetyl is allowed to drop to a low level. By adding the enzyme α-acetolactate decarboxylase (ALDC) at the initial stages of the primary fermentation process, it is possible to circumvent the diacetyl step.

The hazing effect is a quality defect in beer. It is characterized by "cloudiness" in the final product. Laccase can be added to the wort or at the end of the process to remove the polyphenolic compounds that may remain in beer. The polyphenol complexes, formed by laccases, can be separated via filtration and removes probable hazing effect. Laccase also helps in removing excess oxygen in beer and increases the storage life of beer.

6.10   THE TEXTILE INDUSTRY

The textile industry has employed harsh conditions since long; hence, the use of enzymes offers a new perspective. Let us summarize a few enzymes used.

The first enzyme application, as early as 1900s, was the use of barley for the removal of starchy size from woven fabrics. The first microbial amylases were used in the 1950s for the same desizing process. Other enzymes have been introduced for industrial applications such as cellulases, catalases, laccases and pectinases have been employed [23, 24].

Cellulases achieve several purposes that are listed below. They loosen the surface fibers of denims so that mechanical action in a washing machine breaks the surface to remove the indigo dye, revealing the white core of the ring-dyed yarns. The first cellulase products for this application were introduced in the 1980s, and today, most denim garments are "stonewashed" using cellulases. The introduction of cellulases resulted in increased washing capacity for the laundries, and reduced damage to garments and washing machines [23–25].

The small fibers protruding from a fabric render a fuzzy surface, and the gradual entanglement of fibrils results in the formation of pills when a garment is worn and washed. Removal of surface fibrils improves fabric quality, keeping the garment in good form for a longer time. Cellulases have been found to remove these fibrils, thus improving the overall quality and life of the fabrics. This technique is also known as biopolishing of fabrics [23–25].

Catalases: Hydrogen peroxide is used in bleaching of fabrics. Subsequent to this, the excess chemical ought to be removed. To circumvent the use of huge quantities of water to wash the enzymes away, the use of a catalase enzyme, which breaks down hydrogen peroxide into water and molecular oxygen is preferred. The advantage of this process is the end products are safe to the environment and do not disturb the dyeing process. Also, due to high enzyme specificity, the removal of all the hydrogen peroxide ensures automatic inactivation of the enzyme [27].

Alkaline pectinase, which loosens fiber structure by removing pectins between cellulose fibrils and eases the wash-off of waxy impurities, is a key enzyme for a bioscouring process [27].

The group of enzymes called laccases or phenol oxidases possesses the ability to catalyze the oxidation of a wide range of phenolic substances, including indigo. The first commercial use of laccases in the textile industry has been in the denim-washing process, where laccase-mediator systems have been used to reduce backstaining and enhance abrasion levels. Laccases also impart stone finish to denim cloths and can also be used at low concentrations in the bleaching of cotton fabrics [25–27].

Laccases have also been reported to be involved in wool dyeing along with dye precursors and modifiers [25–27]. Chlorination was used traditionally in wool-proof shrinking, which is being replaced by laccases.

Amylases are the second type of enzymes used in the formulation of enzymatic detergent, and many liquid detergents contain these enzymes. These enzymes are used in detergents for laundry and automatic dishwashing to degrade the residues of starchy foods such as potatoes, gravies, custard, chocolate, etc. to dextrins and other smaller oligosaccharides. Removal of starch from surfaces is also important in providing a whiteness benefit, because starch can be an attractant for many types of particulate soils. Examples of amylases used in the detergent industry are derived from *Bacillus* or *Aspergillus* [26, 27].

6.11 ENZYMES IN THE LEATHER INDUSTRY

The various stages in processing of hides to produce leather find applications of enzymes. To make leather pliable, the hides and skins require an enzymatic treatment before tanning know as bating. Here, proteases that work under acidic or alkaline conditions find use [28].

Proteases and *lipases* are used in the soaking step where flesh, blood, etc. are removed from the hides to get a good quality of leather. Unhairing traditionally employs sulfide compounds, and proteases, a more environmentally safe option can be used for this purpose.

Lipases hydrolyze not just the fat on the outside of the hides and skins, but also the fat inside the skin structure. Once most of the natural fat has been removed, subsequent chemical treatments such as tanning, re-tanning and dyeing have a better effect [28].

The main advantages of using lipases are a more uniform color and a cleaner appearance. Lipases also improve the production of hydrophobic (waterproof) leather.

6.12 ENZYMES IN THE COSMETIC INDUSTRY

Papaya and pineapple are the main source of proteolytic (protein dissolving) plant enzymes. Papain is a highly potent plant enzyme isolated from papaya when the fruit is unripe and green. Bromelain is the other proteolytic enzyme obtained from the pineapple plant. The cosmetic industry, however, has for some time now been using the proteolytic (dissolves protein) enzymes such as papain, bromelain, and others for resurfacing and skin smoothing [29].

These substances have proven to be a very useful tool for skin treatment in treating many skin conditions related to skin aging, acne, congestion, and pigmentation.

Super Oxide Dismutase (SOD) in combination with catalase is responsible for protecting the proteins from aging due to oxidation. SOD works by dismutation, a process by which a dangerous highly reactive oxygen free radical is converted to a less reactive from [29].

Diacyl glycerol acyl transferase is used for the renewal of hair and the epidermis. *Lysyl* and *prolyl hydrolases* along with Vitamin C help in maintaining the structure and integrity of collagen.

6.13 ENZYMES AND BIOSURFACTANTS

Biosurfactants (BS) are amphiphilic compounds produced on living surfaces, mostly microbial cell surfaces, or excreted extracellularly and contain hydrophobic and hydrophilic moieties that reduce surface tension. Hydrophobic pollutants present in petroleum hydrocarbons, and soil and water environment require solubilization before being degraded by microbial cells. Mineralization is governed by desorption of hydrocarbons from soil. Surfactants

TABLE 6.2   Some Important Industrial Enzymes and Their Sources [33]

Industrial use	Enzyme: source
Food	Lipase: pancreas, *Rhizopus, Candida*
	Catalase: liver, *Aspergillus*
	Actinidin: Kiwi fruit
	Ficin: latex from figs
	Lipoxygenase: soybean
	Glucose oxidase: *Aspergillus*
	Dextranase: *Penecillium*
	Raffinase: *Mortierella*
Cheese	Rennet: *Mucor, Abomassum*
Dairy	Lactase: *Kluyveromyces; Apergillus*
Brewing	α, and β amylase; β glucanase: malted barley
	Bromelain: latex from pineapple
Tendering of meat	Papain: papaya latex
Baking	α amylase, protease
Confectionary	Invertase: *Saccharomyces*
pharma	Aminoacylase: *Aspergillus*; amidase: *Bacillus*
Waste processing	Cellulase: *Trichoderma*
Detergents	Proteases: *Bacillus*

can increase the surface area of hydrophobic materials, such as pesticides in soil and water environment, thereby increasing their water solubility [30, 31].

Afrouzossadat Hosseini-Abari et al. [32] showed that exopolysaccharide biosurfactant with laccase and catalase enzymes showed potential in degradation of toluene that can find potential in cleaning up oil spills [32] (Table 6.2).

6.14 SUMMARY

In summary, white biotechnology is not an end-of-the-pipe cleaning technology: it is a clean production process that minimizes waste before it is even produced. White biotechnology has substantial potential to reduce environmental impact.

6.15 REVIEW QUESTIONS

1. Briefly describe what is white biotechnology. Can you describe various substances/chemicals that are produced employing white biotechnology?
2. How can bioethanol be produced from sugarcane bagasse?
3. What are bioplastics?
4. How does use of bioplastics help enhance the application of white biotechnology

KEYWORDS

- **bioplastics**
- **biotransformation**
- **brewery**
- **enzyme based biotechnology**
- **white biotechnology**

REFERENCES

1. http://www.dupont.com/products-and-services/fabrics-fibers-nonwovens/fibers/brands/dupont-sorona/articles/how-dupont-sorona-is-made.html.
2. http://www.natureworksllc.com/The-Ingeo-Journey/Eco-Profile-and-LCA/How-Ingeo-is-Made.
3. http://www.henkel.com/sustainability/dialog-and-contacts/positions/white-biotechnology.
4. Frazzetto, G., (2003). White biotechnology. *EMBO Rep.*, *4*(9), 835–837.
5. Ezebuiro, V., Ogugbue, C. J., Oruwari, B., & Ire, F. S., (2015). Bioethanol production by an ethanol-tolerant Bacillus cereus strain GBPS9 using sugarcane bagasse and cassava peels as feedstocks. *J. Biotechnol. Biomater.*, *5*(213), doi:10.4172/2155-952X.1000213.
6. Morgan, S. F., Valentino, F., Hjort, M., Cirne, D., Karabegovic, L., Gerardin, F., Johansson, P., Karlsson, A., Magnusson, P., Alexandersson, T., Bengtsson, S., Majone, M., & Werker, A., (2014). Polyhydroxyalkanoate (PHA) production from sludge and municipal wastewater treatment. *Water Sci. Technol.*, *69*(1), 177–184.
7. Hans-Josef, E., & Andrea, S. R. *Basics of PHA*, Bioplastics Magazine, University of Applied Sciences, Hanover. (3/11) *vol. 6.* Available on:https://f2.hs-hannover.de/fileadmin/media/doc/ifbb/Bioplastics_Magazine__03_11__Vol._6_S._43-45.pdf.
8. Freund, A., (1881). On the formation and preparation of trimethylene alcohol from glycerol. *Monatshefte fur Chemie*, *2*(1), pp. 636–641. doi10.1007 / BF01516545.
9. Kaur, G., Srivastava, A. K., & Chand, S., (2012). Advances in biotechnological production of 1,3-propanediol. *Biochem. Eng. J.*, *64*, 106–118.
10. Da Silva, G. P., Contiero, J., Neto, P. M. A., & Lima, C. J. B., (2014). 1,3-propanediol: production, applications and biotechnological potential. *Chem. Nova., 37*(3). http://www.scielo.br/scielo.php?script=sci_arttext&pid=S0100-40422014000300023
11. Dietz, D., & Zeng, A. P., (2014). Efficient production of 1,3-propanediol from fermentation of crude glycerol with mixed cultures in a simple medium. *Bioprocess Biosyst. Eng., 37*, 225. doi: 10.1007/s00449–013–0989–0.
12. Drozdzynska, A., Leja, K., & Czaczyk, K., (2011). Biotechnological production of 1,3-propanediol from crude glycerol. *Biotechnologia., 92*(1), 92–100.
13. *Industrial Biotechnology*. Nature's own chemical plant: www.igb.fraunhofer.de. https://www.igb.fraunhofer.de/content/dam/igb/en/documents/brochures/_uber/Industrial_Biotechnology__Nature's_own_chemical_plant.pdf.
14. Sadhukhan, S., Villa, R., & Sarkar, U., (2016). Microbial production of succinic acid using crude and purified glycerol from a *Crotalaria juncea* based biorefinery. *Biotechnol. Rep.*, 84–93.
15. Berger, R. G. White biotechnology: Sustainable options for the generation of natural volatile flavours. *Expression of Multidisciplinary Flavour Science. Proceedings of the 12th Weurman Aroma Symposium* (Blank, I., Wust, M., & Yeretzian, C., eds.), pp. 319–327. Winterthur, Switzerland: ZHAW, Zurcher Hochschule fur Angewandte Wissenschaften.

16. Liu, Z. Q., Xiang, Z. W., Shen, Z., Wu, Q., & Lin, X. F., (2014). Enzymatic enanti-oselective aldol reactions of isatin derivatives with cyclic ketones under solvent free conditions. *Biochimie.*, *101*, 156–160. 10.1016/j.biochi.2014.01.006.

17. Li, H. H., He, Y. H., Yuan, Y., & Guan, Z., (2011). Nuclease p1: a new biocatalyst for direct asymmetric aldol reaction under solvent-free conditions. *Green Chem.*, *13*, 185–189. 10.1039/c0gc00486c.

18. Zhou, L. H., Wang, N., Zhang, W., Xie, Z. B., & Yu, X. Q., (2013). Catalytical pro-miscuity of α- amylase: synthesis of 3-substituted 2H-chromene derivatives via bio-catalytic domino oxa-Michael/aldol condensations. *J. Mol. Catal. B. Enzym.*, *91*, 37–43.

19. Gao, N., Chen, Y. L., He, Y. H., & Guan, Z., (2013). Highly efficient and large-scal-able glucoamylase catalyzed Henry reactions. *RSC Adv.*, *3*, 16850–16856. 10.1039/c3ra41287.

20. Werneburg, M., Busch, B., He, J., Richter, M. E. A., Xiang, L., Moore, B. S., Roth, M., Dahse, H. M., & Hertweck, C., (2010). Exploiting enzymatic promiscuity to en-gineer a focused library of highly selective antifungal and anti-proliferative aureothin analogs. *J. Am. Chem. Soc.*, *132*, 10407–10413.

21. Aurora, B, Mukherjee, J., & Gupta, M. N., (2014). Enzyme promiscuity: using the dark side of enzyme specificity in white biotechnology. *Sustain. Chem. Proc., 2*, 25.

22. http://www.biokemi.org/biozoom/issues/522/articles/2368.

23. Sang Yup Lee & Seh Hee Jang, (2006). *White Biotechnology.* Asiabiotech.com. Commentaries. *Asia-Pacific Biotech News, 10*(10), 559, https://doi.org/10.1142/S0219030306000784.

24. *White Biotechnology*: Gateway to a more sustainable future. *EuropaBio.* http://www.europabio.org/industrial-biotech/publications/white-biotechnology-gateway-more-sustainable-future.

25. Boodhoo, K., & Harvey, A., (2013). *Process Intensification for Green Chemistry.* Wiley and Sons Ltd, London. . Edited by KameliaBoodhoo & Adam Harvey, School of Chemical Engineering and Advanced Materials, New Castle University UK.

26. De Souza, P. M., & De Oliveira, M. P., (2010). Application of microbial α-amylase in industry – A review. *Braz. J. Microbiol.*, *41*(4), 850–861.

27. Doshi, R., & Shelke, V., (2001). Enzymes in textile industry- An environment friend-ly approach. *Ind. J. Fib. Text. Res.*, *26*, 202–205.

28. Kamini, N. R., Hemachander, C., Mala, J. G. S., & Puvana krishnan, R. Year of pub-lication 2010 Microbial enzyme technology as an alternative to conventional chemi-cals in leather industry. http://www.iisc.ernet.in/currsci/jul10/articles16.htm.

29. Lods, L. M., Dres, C., Jhonson, C., Scholz, D. B., & Brooks, G. J., (2000). The future of enzymes in cosmetics. *Int. J. Cosmet. Sci., 22*, 85–94.

30. Gnanamani, A., Kavitha, V., Radhakrishnan, N., Sekaran, G., Suseela, G., Rajaku-mar., & Mandal, A. B., (2010). Microbial biosurfactants and hydrolytic enzymes mediates *in situ* development of stable supra-molecular assemblies in fatty acids re-leased from triglycerides. *Colloids and Surfaces B: Biointerfaces*, *78*(2), 200–220.

31. Barros, F. F. C., Simiqueli, A. P. R., De Andrade, C. J., & Pastore, G. M., (2013). Production of enzymes from agroindustrial wastes by biosurfactant-producing strains of bacillus subtilis. *Biotechnol. Res. Int.*, Article ID 103960, http://dx.doi.org/10.1155/2013/103960.

32. Hosseini-Abari, A., et al., (2012). The role of exopolysaccharide, biosurfactant and peroxidase enzymes on toluene degradation by bacteria isolated from marine and wastewater environments. *Jundishapur J. Microbiol.*, 5(3), 479–485. DOI: 10.5812/jjm.3554.

33. http://www1.lsbu.ac.uk/water/enztech/sources.html.

PART II

AGRICULTURAL BIOTECHNOLOGY AND NOVEL AGRICULTURAL PRACTICES

CHAPTER 7

AGRICULTURAL BIOTECHNOLOGY: INTRODUCTION AND HISTORY

PRERNA PANDEY and ANJALI PRIYADARSHINI

CONTENTS

7.1   A BRIEF OVERVIEW

Humans have been engaged in attempts to domesticate nature across ages to help them generate ample food for the growing population. The human society achieved remarkable progress in our comprehension of agricultural biotechnology [1]. A few of the many scientific milestones that have played a key role in agronomic agricultural biotechnology are discussed in the following sections.

7.1.1   PRINCIPLE OF INHERITANCE

Gregor Mendel (1822–1884) was the first person to understand the inheritance of specific traits or characters. He studied this phenomenon first in garden pea (*Pisum sativum*) and stated that the inheritance of specific traits such as flower color, seed shape, and stem length occurred in completely predictable and quantitative ways. He also deduced monohybrid and dihybrid ratios and established quantitative data that the inheritance of characters was in a proportion. His studies spanned research that led to the development of crop breeding techniques and hybrid crops [1, 2].

7.1.2   HYBRID CORN

Henry Wallace and others in the 1920's developed hybrid corn, which was later commercialized in the 1930's. It has several characteristics like drought resistance and is developed by inbreeding technologies [3].

7.1.3   THE HEREDITARY MATERIAL

In 1944, the mystery of the nature of the hereditary material was resolved by Oswald Theodore Avery, Colin MacLeod, and Maclyn McCarty, who stated that the hereditary substance responsible for the transformation of rough pneumococcus bacteria to smooth strain must be the deoxyribonucleic acid (DNA).

7.1.4   DISCOVERY OF JUMPING GENES

Transposons are sections of DNA genes that move from one location to another on a chromosome, hence their name jumping genes. Interestingly, transposons may be manipulated to alter the DNA inside living organisms. Barbara McLintock (1950) discovered an interesting effect of transposons to change the color of kernels of corn.

7.1.5 DEOXYRIBONUCLEIC ACID (DNA)

This was followed by the unraveling of the chemical structure for DNA in 1953 by James D. Watson and Francis Crick. This structure was consistent with the experimental data from diffraction studies and could provide an explanation for DNA replication and transfer of hereditary information from parents to their progeny.

7.1.6 TRIPLET CODE OF THE GENES

Marshall Nirenberg and colleagues in 1961 proposed the language of the genetic code, the three-letter genetic nucleotides present in ribonucleic acid (RNA) and translated into the amino acid sequence of proteins.

7.1.7 PLASMIDS

These extra-nuclear genetic materials were discovered by Stanley Cohen and colleagues in 1973. Their discovery ushered in a new aspect in genetic manipulation studies as plasmids found use as vectors.

7.1.8 TI PLASMID OF PLANTS

In 1977, Mary-Dell Chilton et al. reported that the tumor inducing genes from Ti plasmid of the soil bacterium *Agrobacterium tumefaciens* could be modified by incorporating exogenous gene sequences that could be transferred and expressed in plants.

7.1.9 PLANT CELL TRANSFORMATION

Horsch and colleagues demonstrated in 1984 the ability to regenerate plants by the method of leaf disk transformation.

7.1.10   POLYMERASE CHAIN REACTION (PCR)

The development of the technique by Kary Mullis enabled researchers to amplify a target DNA sequence into exponential copies.

7.1.11   MICROPROJECTILE MEDIATED TRANSFORMATION

The year 1987 saw the emergence of microprojectiles coated with DNA to transform living cells, thus finding use in cell transformation.

7.1.12   INTRODUCTION OF FIRST GENETICALLY MODIFIED FOOD PRODUCT

In 1994, Flavr Savr tomato, the first genetically modified food product, showed delayed ripening and increased shelf life, was introduced. In 1995–1996, two additional crops that were commercialized in USA were Roundup Ready soybeans, designed to be tolerant to the broad- spectrum herbicide Roundup® and YieldGard® corn, which showed resistance against the European corn borer by the presence of a protein from the soil microbe *Bacillus thuringiensis* (Bt).

7.1.13   GOLDEN RICE

The development of "golden rice" by Potrykus and colleagues in 2000 demonstrated an excellent potential for agricultural biotechnology to enhance the nutritional value of food. Due to increased levels of β-carotene, the precursor of vitamin A introduced in the rice genome by recombinant DNA technology, the rice was golden and more nutritional. The U.S. Environmental Protection Agency (USEPA) approved in 2003 the first transgenic root worm-resistant corn that required less quantities of pesticides.

7.2   THE REGULATORY AGENCIES

As agricultural biotechnology deals with food, regulatory programs are essential to the worldwide acceptance of agricultural biotechnology. This demands careful study and evaluation of biotechnological tools before their release into the environment. In the U.S., the following are involved in governmental safety evaluation of the products of agricultural biotechnology.

- *USDA (United States Department of Agriculture):* USDA sets greenhouse standards and inspections. It is involved in field trail authorization and authorization of transport for field trials.
- *EPA (Environmental Protection Agency):* It is involved in approval of species in experimental use permit and the determination of food tolerance or tolerance exemption. It also plays a role in product registration
- *FDA review process*: In general, in the FDA evaluation, the safety of the following aspects in recombinant technology is examined: source organism and genetic material used in the transformation process, the products resulting from the gene expression, and the whole plant or food product derived from the plant.
- *Canadian Food Inspection Agency:* The CFIA promotes food safety awareness through public engagement and verifies industry compliance with standards and science-based regulations.
- *Organisation for Economic Co-operation and Development (OECD):* OECD is a democratic forum of countries committed to the market economy, and coordinating domestic and international policies.
- *European Food Safety Authority (EFSA):* EFSA is involved in scientific risk assessment and communication of its findings to the public.
- *WHO (FAO):* The WHO and FAO published the Codex Alimentarius (Latin for " food code") in 1963, which serves as a beacon and guideline to food safety. It deals with food labeling of nutrients, authorized uses and specifications for food grade chemicals or additives, the tolerance for food contaminants, and codes of hygienic practices concerned with food.

7.3 IMPORTANCE OF AGRICULTURAL BIOTECHNOLOGY IN DEVELOPING COUNTRIES

Despite tremendous growth and development, nearly 800 million people stay hungry, and thousands of hunger-related deaths occur every day. By 2020, the number of undernourished people could well surpass 1 billion [4, 5].

This has augmented a need to propel research in genetically modified crops in order to allow growth of crops to meet global hunger, overcome the curse of malnutrition, and augment the nutritional value of food.

"In developing countries about 650 million of the poorest people live in rural areas where the local production of food is the main economic activity. Without successful agriculture, these people will have neither employment nor the resources they need for a better life. Farming the land is the engine of progress in less developed countries" [4–6].

The development of new plant varieties with higher yields and resistance to abiotic stress (like drought and salinity) and biotic stress factors (like bacteria, fungi, insects) is a vital milestone in addressing the increase in agricultural productivity.

Cassava is an important crop in many countries, from Africa to Indonesia to South America, but 1998 saw a loss of 60% crop to mosaic virus in Africa in 1998. Sweet potatoes are a source of nutrients to the developing countries, but the sweet potato weevil and the feathery mottle virus (SPFMV) have wreaked havoc in Africa: up to 80 % of expected yields have been lost in Africa. The European corn borer destroys approximately 7 %, or 40 million tons, of the world's corn crop every year; this is equivalent to the annual food supply, in calories, for 60 million people.

Hence, biotechnologists are conducting research at a furious pace to address these issues and challenges that cause damage to agriculture. In 1997, the World Bank Consultative Group on International Agricultural Research estimated that biotechnology could help improve world's food production by up to 25 % [4–6].

Cassava is being developed that contains more carotene and minerals. "Golden rice" offers a promise to developing countries. Due to the presence of the beta- carotene gene in rice by recombinant DNA technology, the rice offers augmented nutrition to the starving. Beta-carotene is a natural vitamin, which is converted in the body into vitamin A. An increase in the level of beta-carotene will address the problem of vitamin A deficiency, which causes sight disorders and night blindness, especially in children [6–8].

The Tata Energy Research Institute (TERI), Delhi, jointly with the Michigan State University (MSU) and Monsanto, has launched a research project with support from the US Agency for International Development (USAID), Washington, DC, to enhance the beta-carotene (pro-vitamin A) content in mustard oil, an extensively used cooking medium in India (Tata Energy Research Institute initiates a research project to enhance Vitamin A content of Mustard Oil . Press release, December 7, 2000).

More than a billion people worldwide consume potatoes, but the nutritional quality of potato tubers is not satisfactory as they contain less proteins due to deficiency in amino acids containing sulfur. Biotechnologists are working on the development of "protato," which is a potato with a higher protein content. Chakraborty et al. [8] developed a potato with enhanced nutrition by expression of seed protein, AmA1 (Amaranth Albumin 1) [8]. A 60% increase in total protein content and certain amino acids was also reported [7].

Dr. Florence Muringi Wambugu is the Founder, Director, and the Chief Executive Officer of Africa Harvest Biotech Foundation International (AHBFI). Dr. Wambugu has made significant contributions in research, development, and improved production of maize, pyrethrum, banana, sweet potato, and forestry in Kenya.

Between 1996 and 2000, China's Office of Genetic Engineering Safety Administration approved 251 cases of genetically modified plants, animals, and recombined microorganisms for field trials, environmental releases, or commercialization. In China, more than 90% of the field trials have been directed toward the development of crops with insect and disease resistance. The infestation of a pesticide- resistant bollworm population in the 1980's fueled research in China to develop a genetically modified Bt cotton plant. In 1997, commercial use of Bt cotton was approved and varieties from publicly funded research institutes and private companies from North America became available to farmers.

7.4   STATUS OF HEALTH OF FARMERS, THE ECONOMIC CONTRIBUTION, AND ENVIRONMENTAL IMPACT OF AGRICULTURAL BIOTECHNOLOGY

The use of reduced amount of pesticides has resulted in lesser exposure of the farmers during spraying. The global reduction in pesticide use have seen a reduction in 224 million kg or 6% in the active ingredient in the past 15 years, with a reduction in the Environmental Impact Quotient (EIQ) by more than 15 %. Consequently, reduction in greenhouse gas (GHG) emissions has been reported as fewer quantities of fossil fuels were used to spray pesticides. A reduction in more than 900 million kg CO_2, equivalent to 427,556 average family cars was recorded. In addition, a direct global farm income benefit of $5 billion was reported [9, 10].

Let us examine a few case studies to support the impact of agricultural biotechnology.

7.5   CASE STUDIES

7.5.1   SOUTH AFRICA

Nearly one-tenth of the population is employed in agriculture, which in turn contributes roughly 4% of the GDP. South Africa is one of the first developing countries to commercially approve biotech crops. There has been more than 150% increase in cultivated area, making it the 8th largest acreage worldwide. Several thousand farmers planted Bt and HT white maize for food and Bt and HT yellow maize for animal feed.

A gross margin increase of monetary returns/hectare is observed. There was increase in revenue, yield, harvest labor, and seed cost, while decrease in pesticide cost and pesticide spray labor. This was translated into social implications especially for female farmer- supporting families.

When we look at the impact on research and development, the University of Cape Town was involved in the development of maize resistant to virus and drought [10, 11].

7.5.2 INDIA

Despite the green revolution that spanned agricultural revolution, it is home to the hungriest in the world. Nearly two-third of the population is employed in agriculture, and this agriculture represents more than 20% of the GDP. Following the adoption of biotech crops, there was more than 3-fold increase in agricultural practice till 2016. India was the 5th largest biotech cultivator worldwide employing 2.3 million farmers.

Biotechnology contributed to an increase in the income of farmers by more than US$ 400 million. The yield of Bt cotton nearly doubled against conventional cotton. Looking at the environmental impact, 78% less amount of pesticide was used for Bt cotton against non-Bt cotton in 2002.

The research and development department has seen an increase in activities. Many centers of plant molecular biology were established in 1990s, with ongoing research on the following crops: chickpea, banana, black gram, brassica, cabbage, cauliflower, coffee, cotton, eggplant, maize, muskmelon, mustard/rapeseed, pigeon pea, potato, rice, tobacco, tomato, and wheat [12, 13].

7.5.3   THE PHILIPPINES

Modern biotechnology must be regarded as one of the tools in modernizing Philippine agriculture. A total of 36% of the population is employed in agriculture that contributes to nearly 15% of the GDP. The major crops are banana, cassava, coconut, corn, mango, pineapple, rice, and sugarcane. The adoption of biotechnology in agriculture saw more than 200% increase from the last 10 years, making it one of the largest biotech cultivator worldwide. Farmers have been growing modified crops like Bt, HT, and stacked maize for animal feed.

Research and development spawned studies on drought- tolerant rice, salinity- tolerant rice, late blight resistant potato and tobacco streak virus, resistant groundnut, and sunflower was developed by the Philippine Rice Research Institute. The Institute of Plant Breeding of the University of the Philippines at Los Baños worked on tobacco streak virus- resistant groundnut and sunflower [13].

7.6   FUTURE PERSPECTIVES: THE ROAD AHEAD

The various tools of recombinant DNA technology can be employed in the development of agriculture. For instance, Voytas and Gao [13] have reviewed how the change in DNA sequence albeit by a few nucleotides can create new crops [13].

The research also needs to find expression in terms of profitability of the farmers and have minimum interference with ecology. The teaming of private and public research centers can help in the smooth transition in research and realization of its true potential.

A few vital points that can offer advantages in developing countries include addressing the various regulations in science and associated trade policies. The development of appropriate infrastructure and farm management practices coupled to good access to information can propel useful development in this field.

Another vital factor is the acceptance of the crops and biotechnology by the consumers. According to Curtis et al. [14], there is a general positive response to biotechnology and GM crops keeping in mind the availability and nutritional value of food [14].

Aaron et al. (2016) reported that Indian consumers were willing to consume cisgenic and GM rice, at discounts. A total of 73% were willing to use cisgenic rice, while 76% were willing to use GM rice [15].

Ronald Cantrell of the International Rice Research Institute in the Philippines says: "To still have hunger in our world of abundance is not only unacceptable, it is unforgivable." World hunger is a complex issue, one for which there is no one answer. Yet, while biotechnology may not be the only solution, it can be a valuable tool in the struggle to feed a hungry world.

7.7   REVIEW QUESTIONS

1.   Can you list at least 6 important discoveries in the field of biotechnology in the last century?
2.   Name a few important regulatory agencies around the world that are involved in governmental safety evaluation of the products of agricultural biotechnology.

KEYWORDS

- case studies scientific milestones
- governmental safety evaluation
- India
- Philippines
- South Africa

REFERENCES

1. http://www.rosebudmag.com/truth-squad/gmo-timeline-a-history-of-genetically-modified-foods.
2. "Codex Alimentarius and Food Hygiene" (PDF). Codex alimentarius. Food and Agriculture Organisation of the United Nations. Retrieved 2007–10–15.
3. http://scgo.ca/about/what-is-hybrid-seed-corn/.
4. http://iatp.org/files/Agricultural_Biotechnology_Poverty_Reduction_a.htm.
5. Persley, G. J., & Doyle, J. J., (1999). Biotechnology for developing country: Agriculture problems and opportunities. *Focus, 2,* Brief 1 OF 10 http://ageconsearch.umn.edu/bitstream/44273/2/focus02.pdf.
6. Hall, A., (2005). Capacity development for agricultural biotechnology in developing countries: an innovation systems view of what it is and how to develop it. *Journal of International Development, 17*(5), pp. 611–630.
7. United Nations Department of Economic and Social Affairs/Population Division. (2004). World population prospects: The 2004 revision, vol. III, Analytical Report. Internet: http://www.un.org/esa/population/publications/WPP2004/WPP2004_Vol3_Final/Chapter1.pdf (accessed January 26, 2007).
8. Chakraborty, S., Chakraborty, N., Agrawal, L., Ghosh, S., Narula, K., Shekhar, S., Naik, P. S., Pande, P. C., Chakrborti, S. K., & Datta, A., (2010). Next generation protein-rich potato by expressing a seed protein gene AmA1 as a result of proteome rebalancing in transgenic tuber. doi: 10.1073/pnas.1006265107.
9. United Nations Millennium Project Task Force on Hunger. (2005), Halving hunger: It can be done, http://www.unmillenniumproject.org/documents/Hunger-lowrescomplete.pdf (accessed 1/27/07).
10. Wambugu, F., (2003). Development and transfer of genetically modified virus-resistant sweet potato for subsistence farmers in Kenya. *Nutrition Reviews, 61*(6), 110–113.
11. World Bank. Agriculture and Rural Development. http://web.worldbank.org/WBSITE/EXTERNAL/TOPICS/EXTARD/0,m enuPK:336688~pagePK:149018~piPK:149093~theSitePK:336682,00. html (accessed 1/7/07).
12. World Health Organization. *Health and Development.* Internet: http://www.who.int/hdp/en/ (accessed 1/7/07).
13. Voytas, D. F., & Gao, C., (2014). Precision genome engineering and agriculture: Opportunities and regulatory challenges. *PLoS Biol., 12*(6), e1001877. doi:10.1371/journal.pbio.1001877.
14. Kynda, R., Curtis, J., McCluskey, J., & Wahl, T. I., (2004). Consumer acceptance of genetically modified food products in the developing world. *AgBiuo Forum., 7*(1–2), Article 13.
15. Shew, A. M., Nalley, L. L., Danforth, D. M., Dixon, B. L., Nayga, Jr, R. M., Delwaide, A. C., & Valent, B., (2016). Are all GMOs the same? Consumer acceptance of cisgenic rice in India. *Plant Biotechnology Journal, 14*(1), 4–7.

CHAPTER 8

PLANT BREEDING AND SEED TECHNOLOGY

PRERNA PANDEY and ANJALI PRIYADARSHINI

CONTENTS

8.1 PLANT BREEDING: INTRODUCTION

Plant breeding is a branch of biology that serves to alter the genotype of plant so that they possess desirable traits. Plant breeding can be defined as a technology of developing superior crop plants or varieties for various purposes, or, as the genetic adjustment of plants to serve mankind.

Plant breeding is an integrated approach that requires an amalgamation of genetics, cytology, morphology, taxonomy, physiology, pathology, entomology, agronomy, soil science, biochemistry, statistics, and bio-metrics. Thus, we may summarize that plant breeding is a technology of developing genetically superior plants in terms of utility to mankind.

8.1.1 OBJECTIVES OF BREEDING

Breeders seek to enhance the following few traits in plant varieties: higher yield, improved quality, disease and pest resistance, wide adaptability, and resistance to various abiotic stress like moisture, salinity, and biotic stress. The rich diversity and variation in crops we see may be environmental or genetic. Environmental variation is dependent on the environment, for instance, soil quality. Genetic or genotype variation refers to the genetic makeup that is independent of the environment, for example, flower color or height in garden pea.

8.2 PLANT BREEDING METHODOLOGIES

8.2.1 INTRODUCTION METHOD

This method is a rapid method of introducing species in a new environment. It is instrumental in introducing new germplasm in new areas and protects the genetic variability. But, one must take into account that this method may cause the introduction of potential weeds like *Parthenium argantatum*, water hyacinth, or certain diseases like late blight of potato from Europe in 1883 and pests like potato tuber moth.

8.2.2 METHODS OF PLANT BREEDING IN SELF-POLLINATED PLANTS

8.2.2.1 Selection

This is an ancient method of crop improvement. Before domestication, crop species were subjected to natural selection. But, the advent of new technology ushered in artificial selection. Selection is dependent on the premises that it is effective for heritable variations and the variation must pre-exist in the population. Self-pollinated crops may be bred by the following two methods:

8.2.2.1.1 Pure line selection

The Danish botanist Johan heralded this technique on the basis of his studies on Princess beans (*Phaseolus vulgaris*). Due to heterogeneity in seed size, he bred the varieties with large and small seeds individually and observed that the progenies of larger seeds are generally larger than those obtained from smaller seeds. This established a genetic basis for the seed size.

The original seed lot was a mixture of pure lines; he proposed that a population of self-fertilized species consists of several homozygous genotypes. Variation in such a population has genetic base, and therefore, selection is effective.

A pure line is a progeny of single self- fertilized homozygous plant. In pure line selection, large numbers of plants are selected from a self-pollinated crop and is harvested individually, individual plant progenies from them are evaluated separately, and the best one is released as pure line variety. Therefore, it is also known as individual plant selection.

All plants in a pure line have the same genotype and the phenotype variations are a result of environmental variation. A superior pure line is used as a parent in the development of a new variety by hybridization as well as for studying mutations and other biological investigations such as medicine, immunology, physiology, and biochemistry.

This method offers few advantages like improvement over the original variety, easy method requiring less skill, and can be used for developing pure lines.

However, the flip side of the coin is that stability and adaptability issues and maintenance of the lines and its practice in cross pollinated crops offer challenges.

A few notable landmarks of this technique have been the development of shining mung –1 selected from Kulu type-1; 202; Kalyan sona from CIMMYT; many developed wheat varieties including NP-4, NP-6, NP-12, NP-52 and NP-28; Mung Var; T-1; B-1; and tobacco chatham special-9.

8.2.2.1.2 Mass selection

It entails a process where plants of similar phenotype are selected and their seeds are harvested and mixed together to form a new variety. This method is practiced in both self- and cross –pollinated crops. The method relies on the phenotype and is practiced for distinct phenotypic traits such as color, height, etc.

This method may have been developed from a pure line homozygous parent that eventually becomes a variant population.

The following two methods are used:

- Hallets Method (1869): the crop is grown under the best environmental conditions followed by mass selection
- Rimpar Method (1867): the crop is grown under ordinary condition or unfavorable conditions and followed by mass selection.

The mass breeding technique employs the following protocol: Numerous phenotypically similar plants are selected at the time of harvest on the basis of their vigor, plant type, disease resistance, and other desirable characteristics. These are harvested and seeds are mixed together to form a composite batch. This composite seed is subjected to preliminary yield trial to select the superior strains. The superior strains are evaluated for their performance in coordinated yield trails at several locations, first in an initial evaluation trail (IET) for one year, and then if found promising, they are promoted to uniform variety trail (UVT) for two or more years. The promising strain may then be released for cultivation.

This technique shows merits like wide adaptability, ease of use, and rapidity that can meet immediate requirements of markets and farmers.

However, the other side of the coin show certain challenges such as nonuniformity and inability to distinguish whether a trait is the result of genotype or environmental variations.

8.2.2.2   Hybridization

The mating or crossing of two plants of dissimilar genotype is known as hybridization. A hybrid is a F1 generation of mating between genetically dissimilar plant. The seed as well as the progeny resulting from the hybridization are known as hybrid of F1. The progeny of F1 obtained by selfing or intermating of F1 and the subsequent generation are termed as segregating generation.

8.2.2.2.1 Hybrid

The progeny of a cross between genetically different plants is called hybrid. Most of the hybrid varieties are F1 from two or more purelines (Tomato, L esculentum) or inbreds (Maize, Zeamays).

Hybrids may be Intraspecific (developed by crossing the same species) or interspecific (F1 progeny between two different progenies of the same genus).

Intraspecific hybrids may be of the following types:

- **Single cross hybrid**: It is a cross of two varieties (for instance, F × G). In cross- pollinated crops, such hybrids are developed from a cross between two inbreds, whereas in self-pollinated crop, they developed from a cross between two homozygous varieties. It shows the maximum amount of heterosis.
- **Double cross hybrid:** It is a cross of two single cross hybrids (for instance, F × G crossed with L × M) [1].
- **Three-way cross hybrids:** These have also been reported. They are an outcome from the cross between an F1 hybrid and an inbred line [2]. Ashakina, et al., [3] studied these hybrids in tomatoes and reported that the highest significant positive heterobeltosis were the single cross hybrids, while the highest significant positive heterobeltosis was found in double cross hybrid for shelf life [3].
- **Interspecific hybrids**: are seldom employed in commercial cultivation due to low fertility. In cotton interspecific hybrids between tetraploid cultivated sp. (*Gossypium hirsutum × G. barbadense*) and diploid cultivated species (*G. arboretum × G. herbaceum*). Example: A tetraploid level: Var. Laxmi, Surti, HB224, etc., and at diploid level: DH-7, DH-9, Pha46, etc.

In cross-pollinated plants, the hybridization process employs few steps as follows:

- **Inbred development**: Plants exhibiting superior characteristics are subjected to selfing repeatedly to develop a pure line. Following the development of the inbred line, it is evaluated. This evaluation may be done by:

- **Top cross method:** The top cross method entails a cross of single inbred lines with a "tester" cross pollinated variety. The lines that produce high- yielding single cross with the tester are selected.
- **Single cross method:** In the single-cross method, selected lines are crossed in all possible combinations. Once superior lines are established, hybrid seeds may be produced. There is a separate section on hybrid seeds in this book.

8.2.2.3   Pedigree Methods

The pedigree may be defined as a description of the ancestors of an individual. In the pedigree method, individual plants are selected from F2 and the subsequent generation and their progenies are tested. During the entire process, a record of the entire parent's offspring relationship is kept, which is known as pedigree record.

The selection of individual plant is continued till the progenies show no segregation. At this stage, selection is done among the progenies, because there would be no genetic variation within progenies.

The procedure of pedigree method involves crossing or hybridization of selected parents. Once, the F1 is produced, its production is optimized to produce maximum number of F2. These are successively developed into the F3, F4, F5, and F6 until an F11 generation.

The superior lines are tested in replicated yield trials at several locations for desirable characteristics. This method is well suited for the improvement of characteristics, achievement of homozygosity, and recovery of characteristics. However, the method is time consuming, the cost of a pedigree record is high, and skill is required.

8.2.2.3.1   Bulk Population Method

The method, also known as mass method, involves growing a species grown in bulk from F1 to F5, while selection is done from F6 or subsequent generations.

Following bulk planting of the F1 until F5, selection on the basis of characteristics is done from subsequent generations on the basis of phenotype. Successive progenies are evaluated for desirable characteristics. Quality tests are carried out at F9 to select superior breeds.

The method has merits like ease, low cost, and elimination of undesirable characteristics. However, the method is time consuming and lacks pedigree knowledge.

A modification of the bulk multiplication is the single seed descent technique. Here, a single seed is isolated from the F2 and bulked into the F3 till the F6. Subsequently, individual plants are subjected to isolation and selection.

8.2.2.3.2 Multiple Crossing

It is a complex system of crossing in which 8 to 16 parents are systematically crossed to develop new hybrid varieties, for example, barley. Multiple crosses are produced by crossing pairs of parents, pairs of F1, and pairs of F1, till all parents are included into a common progeny.

This method of breeding help to accumulate more and more gene quickly from several parents. But sometimes, undesirable combinations may be brought together, because large number of parents are involved. This can be avoided by selecting the desirable combination before including in crossing program during each generation. These procedures will require a longer time to reach the final cross. Self-pollinated crops like wheat and rice are common examples here.

8.2.2.3.3 Back Cross Method

A cross between F1 hybrid and one of its parents is known as a backcross. In this method, two plants are selected as parents and crossed. The hybrid so produced is successively backcrossed to one of their parents. Consequently, hybrid backcross progeny become increasingly similar to that of the parent.

Backcross breeding enables breeders to transfer a desired trait such as a transgene from one variety (donor parent, DP) into the favored genetic background of another (recurrent parent, RP).

If the trait of interest is produced by a dominant gene, this process involves four rounds of backcrossing within seven seasons. In the case of a recessive characteristic, this process requires more generations of selfing, and thus, nine or more seasons are needed [4].

The method necessitates a parent that lacks a characteristic or two, while the characteristics under study must be heritable. Backcrosses are commonly employed in the transfer of resistance characteristics and seed characteristics either Intra- or Intervaraietal.

8.2.3 CROSS- POLLINATED PLANTS MAY BE BRED BY THE FOLLOWING TECHNIQUES

8.2.3.1   Progeny Selection (Ear to Row Method)

This method was developed in 1908 by Hopkins and is extensively used in maize. Around 50 to 100 plants are selected for phenotype and subjected to cross pollination. The resultant seeds are harvested and grown from selected plants. Successive rounds of plants are selected for their superior characteristics.

This method is relatively simple and the selection cycle is of one year only. However, it suffers from the defect that the weak and inferior progenies pollinate plants in the superior progenies. This reduces the effectiveness of selection.

8.2.3.2   Plant to Row Method

Louis De Vilmorin developed this technique for sugarbeet. In this method, plants are selected, subjected to cross- pollination, and the individual plants are harvested separately. Superior progenies are collected for desirable characteristics.

8.2.3.3   Male Sterility

Male sterility is defined as an absence or non-function of pollen grain in plant or incapability of plants to produce or release functional pollen grains. Cytoplasmic genetic male sterility is widely used for hybrid seed production of both seed propagated species and vegetatively propagated species. It is used commercially to produce hybrid seed in maize, pearl millet, cotton, rice, sunflower, and jowar. The use of male sterility in hybrid seed production is of great value and has been discussed in the section of hybrid seeds (Section "Production of hybrid seeds")

8.2.3.4   Heterosis

The superiority of F1 hybrid over both its parents in terms of yield or some other characteristics is heterosis [5].

In the words of G. H. Shull, "The physiological vigor of an organism as manifested in its rapidity of growth, its height and general robustness, is positively correlated with the degree of dissimilarity in the gametes by whose union the organism was formed. The more numerous the differences between the uniting gametes — at least within certain limits — the greater on the whole is the amount of stimulation. These differences need not be Mendelian in their inheritance. To avoid the implication that all the genotypic differences which stimulate cell-division, growth and other physiological activities of an organism are Mendelian in their inheritance and also to gain brevity of expression I suggest that the word 'heterosis' be adopted" [5].

The molecular mechanisms underlying heterosis are complex. Schnable and Springer [6] suggest that multigene models may explain the phenomenon [6]. Complementation of allelic variation, genes, and gene expression along with epigenetics have all been hinted at to suggest explanations of the phenomenon.

8.2.4 BREEDING METHODS IN ASEXUALLY PROPAGATED CROP

There are some agricultural (sugarcane, potato, sweet potato, etc.) and horticultural (banana, mango, citrus, pears, peaches, litchi, etc.) crops that propagate by asexual means. Such asexually reproducing crops are of nutritional value: The Food and Agriculture Organization ranks potato as the fourth crop after wheat, maize and rice, and cassava as the sixth crop after barley [7].

For such plants, "clonal selection" is employed. Clone is defined as a genetically uniform material derived from a single individual that is vegetatively propagated either *in vivo* or *in vitro*. Plants are selected having desirable characteristics from a population. The selected plants are propagated as clones that are routinely examined with elimination of inferior clones. These clones are then subjected to replicated yield trials, following which they may be released as a new variety.

8.2.4.1 Hybridization

Improvement in such clonal crops may be done by hybridization of desirable clones. Clones to be used as parents are grown and crosses are made to produce F1 progeny. This is followed up by successive testing of superior clones and discarding of the inferior ones.

Dilson Antônio Bisognin [7] reported the importance of asexually reproducing crops like potato, cassava, yams, and sweet potatoes [7]. The author suggests the possibility of combining asexual and sexual methods of reproduction to develop new cultivars. The author recommends the use of biotechnology tools like markers to identify superior plants at the first step in breeding programs.

8.3 CONCLUDING REMARKS: THE ROAD AHEAD

Jiangfeng, He, et al., [8] suggest the use of genotyping-by-sequencing (GBS) as marker- assisted sequencing [8]. They suggest its use in marker-assisted techniques of plant breeding.

GBS is a straightforward method for small genomes. However, target enrichment must be employed to ensure sufficient overlap in larger genome sequence coverage, which could be technologically challenging, time-consuming, and costly for assaying large numbers of samples (Figure 8.1).

Reducing genome complexity with restriction enzymes (REs) can avoid repetitive regions of genomes , and lower copy regions can be targeted with two to three-fold higher efficiency, which tremendously simplifies computationally challenging alignment problems in species with high levels of genetic diversity [9].

Cooper, et al., [10] reviewed that the scaling up of qualitative biology shall continue to assist plant breeding. This would require an integration of quantitative genetics, statistics, and gene-to-phenotype knowledge [9].

8.4   SEED TECHNOLOGY: HYBRID SEEDS

Seed may be defined as a mature integumented megasporangium (ovule) composed of embryo(s) along with storage food material covered by a protective coat.

The term "hybrid," refers to a plant variety developed through a cross of two genetically dissimilar parent plants. Hybrids are formed naturally spontaneously and randomly, when open-pollinated plants cross-pollinate

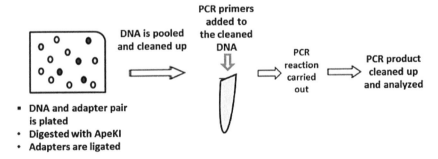

FIGURE 8.1   Steps in GBS library construction, as demonstrated by Elshire, et al., [9]. Up to 96 DNA samples can be processed simultaneously. (1) DNA samples, barcode, and common adapter pairs are plated and dried; (2–3) samples are then digested with *Ape*KI and adapters are ligated to the ends of genomic DNA fragments; (4) T4 ligase is inactivated by heating and an aliquot of each sample is pooled and applied to a size exclusion column to remove unreacted adapters; (5) appropriate primers with binding sites on the ligated adapters are added and PCR is performed to increase the fragment pool; (6–7) PCR products are cleaned up and fragment sizes of the resulting library are checked on a DNA analyzer. Libraries without adapter dimers are retained for DNA sequencing [9].

with other related varieties. For creating hybrid seeds, plant breeders exert control over the process.

The advantage of production of hybrid seeds compared to inbred, open-pollinated lines comes from the ability to cross the DNA of two *different yet related plants* to produce new, desirable traits that otherwise cannot be achieved through inbreeding the *same* plants.

Hybridization has opened a vast repertoire of varieties in the market, including canola, grapefruit, sweet corn, cantaloupes, seedless watermelons, "burpless" cucumbers, tangelos, clementines, apriums, pluots, to name a few.

Both plums and apricots come from the same genus—*Prunus*—which made crossing the two fruits relatively easy for Zaiger to develop pluots [11]. Plumcots are plum-apricot hybrids. Tangelos are a cross between tangerines and grapefruits created by USDA scientist named Walter Swingle in 1911. A cross between a lime and a kumquat is a Limquat.

Mandarins and papedas of East Asia were crossed to produce a Yuzu. Hybrid lilies are classified as Asiatic hybrids and Oriental hybrids. Olympia is a hybrid of spinach, which is preferred due to its superior growth patterns. Meyer lemons, originating in China, are a cross between a true lemon tree and mandarin orange tree that are sweeter to taste. Early sunglow is a hybrid of sweet corn that is sweeter while quick and easy to harvest. Better Boys tomatoes are hybrids that are resistant to a number of pests and diseases like erticillium wilt, fusarium wilt, and nematodes.

The technique of hybridization enables us to "tailor make" varieties like that of drought or pest resistance. The advent of hybrid seeds started with maize, and now, it has extended to many plant families. The requisites for hybrid seed production in parental lines include good seed and pollen production in the female and male plants, respectively.

The basic procedures for hybrid seed production include development and identification of parental lines, multiplication of parental lines, and crossing between parental lines and production of F1 generation.

8.5 PRODUCTION OF HYBRID SEEDS

Pollen from male parent is used to pollinate and fertilize ova in a female parent to produce F1 hybrid seeds. Naturally, hybrids are produced by cross- pollination, and certain mechanisms have been studied that ensure the success of cross- pollination. Let us see a few of these in brief [12, 13]:

8.5.1 DICLINY

Dicliny is also known as unisexuality. In monoecious plants, male and female flowers are borne on the same plant, as in maize. In dioecious plants, male flowers are borne on different plants, as in papaya.

8.5.2 DICHOGAMY

Here, the time of dehiscence of the anthers and ovary is different, thus ensuring cross- pollination.

8.5.3 SELF-INCOMPATIBILITY

In certain species such as petunia, self-fertilization is avoided by recognizing the pollen from the same species by the stigma.

8.5.4 HERKOGAMY

Herkogamy involves difference in the spatial arrangement in the stigma and anther.

8.5.5 MALE STERILITY

Absence of or malformed male sex organ (functional pollen) in normal bisexual flower. Male sterility is of three types: genetic male sterility, cytoplasm sterility, and cytoplasmic- genetic male sterility, which would be elaborated in a subsequent section.

8.6 DESIRABLE FEATURES OF PARENTAL LINES

a. Female parent should have a high yield of seeds with good characteristics and presence of male sterility.
b. Male parent should produce copious amounts of pollen.

8.7   BASIC PROCEDURES FOR HYBRID SEED PRODUCTION

- Development and identification for parental lines.
- Multiplication of parental lines.
- Crossing between parental lines and production of F1.

Few crossing mechanisms have been adopted for commercial hybrid seed production that include:

8.7.1   EMASCULATION

Hybrid seeds are produced manually by removal of the anthers from female parent before the formation or release of pollen. This system is feasible when the anther and carpels are present in a single flower or sexes are separate. Following the removal of anther, the stigma of the female parent is exposed to pollen of the desired male parent. In bisexual flowers, emasculation is essential to prevent self-pollination. For species with large flowers, removal of anthers is possible with the help of forceps. It is done before anther dehiscence. In plants with small pollen, the gentle suction method, in which a thin rubber or a glass tube is attached to a suction hose, is used to suck the anthers from the flowers.

8.7.2   SELF-INCOMPATIBILITY

It refers to the failure of pollen to fertilize the same flower or other flower of the same plant, or the failure of pollen tube to penetrate the full length of style and achieve fertilization. Two self-incompatible, but cross –compatible, lines are inter planted, and seeds obtained from both the lines would be a hybrid seed. Alternatively, a self-incompatible line may be inter planted with a self-compatible line. From this scheme, seed from only the self –incompatible line would be hybrid.

8.7.3   MALE STERILITY

Male sterility is defined as an absence or nonfunction of pollen grain in plant or incapability of plants to produce or release functional pollen grains. The use of male sterility in hybrid seed production has a great importance as it eliminates the process of mechanical emasculation.

Pollen sterility, which is caused by nuclear genes, is termed as genic or genetic male sterility (GMS). The genes affecting stamen and pollen development are affected. The male sterility alleles may develop spontaneously, or they can be induced artificially. GMS is usually recessive and monogenic and hence can be used in hybrid seed production. It is used in both seed- propagated crops and vegetatively propagated species. A male sterile line is maintained by crossing with heterozygous male fertile line. Male sterile plants of monoecious or hermaphrodite crops are potentially useful in hybrid program because they eliminate the labor-intensive process of flower emasculation.

8.7.4   CYTOPLASMIC MALE STERILITY (CMS)

Pollen sterility that is controlled by cytoplasmic genes is known as cytoplasmic male sterility (CMS). This finds application in the propagation of ornamental plants. They show non-Mendelian inheritance and are under the control of cytoplasmic factors. It is influenced by extra nuclear genome (mitochondria or chloroplast) and shows maternal inheritance

When pollen sterility is controlled by both cytoplasmic and nuclear genes, it is known as cytoplasmic genetic male sterility. The sterility may be under control of extranuclear genome and show nuclear restorations of the fertility genes, hence showing a combination of both genetic and cytoplasmic factors.

Cytoplasmic genetic male sterility is widely used for hybrid seed production of both seed- propagated species and vegetatively propagated species. Incorporation of these systems for male sterility evades the need for emasculation in cross-pollinated species, thus encouraging cross- breeding and producing only hybrid seeds under natural conditions.

Male sterile plants may be produced by emasculation or by identifying certain lines that do not produce viable pollen.

These crossing processes will result in heterosis [14]. Heterosis, also known as "hybrid vigor," refers to the phenomenon wherein progeny of diverse varieties of a species or crosses between species exhibit greater biomass, speed of development, and fertility than both parents.

The molecular bases for this phenomenon remain elusive. Recent studies in hybrids and allopolyploids using several approaches like transcriptomic, proteomic, metabolomic, epigenomic, and systems biology have provided new insights. It is likely that heterosis arises from allelic interactions between parental genomes, leading to altered programming of genes that promote the growth, stress tolerance and fitness of hybrids. For example, epigenetic modifications of key regulatory genes in hybrids and allopolyploids can alter complex regulatory networks of physiology and metabolism, thus modulating biomass and leading to heterosis [14, 15].

8.7.5   CASE STUDIES IN HYBRID SEEDS

Currently, commercial hybrid rice production includes a CMS line-based three-line system and a photoperiod/ thermo-sensitive genic male sterile (PTGMS) line-based two-line system [16]. To overcome the intrinsic drawbacks involved, Chan et al. [15] developed a rice male sterility system involving transgenic technology and found it to be promising in the field of hybrid seed technology [15–17].

Singh et al. [18] give a comprehensive summary of the merits and demerits of hybrid seeds using maize [18]. The advantages of growing hybrid maize include the following: hybrids are generally higher yielding than open- pollinated varieties, they express uniformity in characteristics like color and maturity, and this uniformity can offer economic advantages.

A few demerits include cost factors and the heterosis factor as the F1 would be different from the parents.

8.8   SEED TECHNOLOGY

Seed technology is a method to improve the genetic and physical characteristics of seeds. It involves processes such as variety development, evaluation and activities related to seeds, like, release production, processing, storage, testing, certification, quality control, and marketing.

Objectives of seed technology include rapid multiplication, prompt and timely supply, and high quality seed at reasonable prices.

8.8.1   GENETIC PURITY DURING SEED PRODUCTION

For the maintenance of genetic purity, various steps include providing adequate isolation, rouging of fields, routine testing of crops, growing crops in their areas of adaptation to prevent genetic shifts, and certification of crops to ensure quality of seeds.

8.8.2   PRINCIPLES OF QUALITY SEED PRODUCTION

Maintenance of genetic purity of seed cell lines is of utmost importance. To avert genetic deterioration of seeds, we must understand a few causes of genetic deterioration that include:

Mechanical mixing: occurs when two different varieties are sown together or the use of threshing machines or bags or storage bins that have contamination with other species. This mixing can be prevented by isolation methods and preventing reuse of storage equipment.

Natural crossing or contamination: occurs in sexually reproducing plants where pollination occurs with undesirable or diseased species. This can be circumvented by isolation mechanisms, especially for wind- or insect- pollinated plants.

Additionally, the prevention of *attack by parasites or pathogens* is essential for genetic purity.

8.8.3   THE ISOLATION METHODS

These are essential to prevent natural contamination or mechanical mixing. This can be followed in a field by adopting isolation distances when two different crops are grown together. It includes space isolation where there is appropriate space between other species or contaminants and a seed field, and time isolation where the flowering of contaminant and seed field should not coincide with each other [19].

8.8.4   ROGUEING

Rogueing includes removal of rogue or undesirable plants. The undesirable plants include other volunteer plants of a previous crop, other crop plants, diseased plants, and weeds. This is an absolute responsibility of a seed grower to prevent contamination for cross- pollinated crops, thus ensuring crop purity.

8.8.5 CONTROL OF SEED SOURCES

Prior to raising a crop, it is essential that the seeds should be obtained from an approved source. There are classes of seeds according to the requirement, and these classes are recognized by AOSCA (Association of Official Seed Certifying Agencies). A buyer can be protected by the seed regulation laws or acts in many countries; for this purpose, in many countries, seed agencies are associated with ISTA (International Seed Testing Association) [19, 20].

- **Breeder's seed:** is a progeny of a nucleus seed that was pure genetically under isolation. They give rise to the foundation seeds.
- **Foundation seed:** is a seed stock so as to maintain specific genetic identity and purity whose production must be carefully supervised by authorized representatives. Foundation seed is the source of all other certified seed classes.
- **Registered seed:** Registered seed is the progeny of foundation seeds, and it is handled so as to maintain genetic identity and purity. It is vital that it is approved by and certified by appropriate certifying agencies.
- **Certified seed:** This is the progeny of the above classes and requires certification from seed certifying organizations to ensure genetic purity. The certification of seeds is a legal and scientific system to ensure protection of genetic purity.
- **Seed certification:** The genetic purity in commercial seed production is maintained through a system of seed certification. The objective of seed certification is to maintain and make available crop seeds, tubers, bulbs, etc., which are of good seeding value and true to variety. For seed certification purpose, well- experienced and qualified personal from seed certification agency are required, and they carry out field inspection at the appropriate stage of crop growth. They also make seed inspection variety the requisite genetic purity and quality.

Periodic tests for genetic purity are important to ensure pure stock of seeds by appropriate agencies.

A few well-established guidelines can serve for seed testing agencies, like responsible and skilled staff, good service, and research activities to augment the seed testing practices [21, 22].

8.8.6   PLANT BREEDER'S RIGHTS (PBR)

Plant breeder's right (PBR) are granted by the government to the breeder, originator, or owner of a plant variety, which empower the breeder to exclude others from producing or commercializing the propagating material of the protected variety for a period of at least 15–20 years.

8.8.7   PRECAUTIONS TO BE FOLLOWED DURING SEED HANDLING

As several of the products applied on seeds are harmful to the ecology, care must be exercised to prevent human or plant or animal consumption and damage. A seed must be treated with appropriate chemical agent to extend its life in the correct dosage. Care must also be taken to ensure water content of the seed is adequate [21, 22].

Seeds may be dried by wagon drying that includes the use of heated air to dry seeds followed by cooling, bag drying where seeds are dried in sacks or bags, and box drying that employs boxes made of local materials to minimize the cost.

Seed Storage: It is vital to maintain a seed in good physical and physiological condition from the time they are harvested until the time they are sown in a field. The kind and variety of seeds and their quality affect the longevity of seeds in storage.

Moisture content: Another important factor is moisture content of the seeds. The following data show moisture content and storage life of cereal seeds at temperatures not exceeding 90 F for seeds of high germination and high vigor at the start of storage [19–21].

Seed Moisture Content Percent	Storage Life
11 to 13	½ Year
10 to 12	One Year
9 to 11	Two Years
8 to 10	Four Years

Seeds stored at high moisture content demonstrate increased respiration, heating, and fungal invasion, resulting in reduced seed vigor and viability, while lower moisture content may cause desiccation.

Humidity: Another important factor in storage is humidity. Temperature also plays an important role in storage, as the number of insect and molds increases as temperature increases. The higher the moisture content of the seeds, the more they are adversely affected by temperature. Decreasing temperature and seed moisture, therefore, is an effective means of maintaining seed quality in storage.

The deterioration of seed quality, vigor, and viability, due to high relative humidity and high temperature during the post-maturation and pre-harvest period is referred to as field weathering. Exposure to hot and humid conditions, rainfall, and photoperiod after ripening are pre-harvest that cause seed quality loss following physiological maturity. Increase in pressure of oxygen tend to decrease the period of viability of seeds.

Various organisms associated with seed: One must also take into account various organisms associated with seed storage . Bacteria and fungi constitute the seed microflora. Research shows a positive correlation between fungi and humidity. The use of appropriate agents is still warranted and under investigation, while deep freezing offers a solution. Insects and mites can be controlled by chemical agents such as methyl bromide, hydrogen cyanide, phosphine, ethylene dichloride, and carbon tetrachloride in 1: 3:1 mixture with carbon disulfide and naphthalene.

The storage sites of seeds should be sealed or screens must be employed to prevent the entry of birds. Additionally, rodent control measures must be taken into consideration to prevent loss of seeds [20].

8.8.8   SEED MARKETING

The selling of packaged seeds along with delivery and sales activity compromise this section. A few pointers include:

8.8.9   FORECASTS OF DEMANDS

It entails that the seed supply keeps pace with seed demand in terms of quantity, quality, price, place, and time. The total cultivation, improved production techniques, competitor study, and climate of growing the crop must be taken into account. Varietal purity and the yield potential of high quality seed of self-pollinating varieties can be maintained by farmers during the reproduction process, without significant deterioration for three to four generations. Appropriate training can be imparted to farmers for this purpose. Periodic surveys of the market must be carried out.

8.8.10   A REALISTIC ASSESSMENT AND TARGETS OF SEED DEMAND

These are very necessary. Excessive quantities result in large carry-over stocks and subsequent losses, On the other hand, short supplies would deprive the seed company. A good marketing structure is vital to achieve good results.

8.8.11   MARKETING STRUCTURE (ESTABLISHMENT OF EFFECTIVE CHANNEL FOR SEED DISTRIBUTION)

The key to success in seed marketing is the establishment of effective channel of distribution. The various channels through which seed can be marketed vary greatly according to the needs of the seed company. A central marketing cell and offices in end- use areas can be used. Sale could be performed by dealers such as private dealers, cooperatives, and agro-sales service centers. The central marketing cell is responsible for planning, appointment of dealers/ distributors, seed movement, market intelligence research, pricing, promotional activities, financing, and record keeping [20–22].

Key factors affecting seed marketing include a clear-cut policy for the seed industry and the availability and development of high yielding varieties that regularly make a successful appearance in the market.

8.9   SUMMARY

Plant breeding is an integrated approach that requires an amalgamation of genetics, cytology, morphology, taxonomy, physiology, pathology, entomology, agronomy, soil science, biochemistry, statistics, and biometrics. Breeders seek to enhance the following few traits in plant varieties: higher yield, improved quality, disease and pest resistance, wide adaptability, and resistance to various abiotic stress like moisture, salinity, and biotic stress. Genotyping by sequencing (GBS) as marker- assisted sequencing is a straightforward method for small genomes. The scaling up of qualitative biology would require an integration of quantitative genetics, statistics, and gene-to-phenotype knowledge and shall continue to assist plant breeding. The advantage of production of hybrid seeds compared to inbred, open-pollinated lines comes from the ability to cross the DNA of two *different yet related plants* to produce new, desirable traits that otherwise cannot be achieved through inbreeding the *same* plants. The basic procedures for hybrid seed production include development and identification of parental lines, multiplication of parental lines, and crossing between parental lines and production of F1 generation.

8.10 REVIEW QUESTIONS

1. Describe the methods of breeding in self-pollinated plants.
2. What is a hybrid plant? Describe how a hybrid of wheat was generated.
3. Describe the methods of breeding in cross-pollinated plants.
4. What is heterosis?
5. How can genotyping-by-sequencing (GBS) be used as marker-assisted technique in plant breeding?

KEYWORDS

- **genotyping-by-sequencing**
- **heterosis**
- **hybrid seed**
- **plant breeder**
- **seed marketing**
- **seed technology**

REFERENCES

1. Rawlings, J. O., & Cockerham, C. C., (1962). Analysis of double cross hybrid populations. *Biometrics, 18*(2), 229–244. doi: 10.2307/2527461.

2. Darbeshwar, R., (2000). *Plant Breeding Analysis and Exploitation of Variation*. Pangbourne, UK, Alpha Science International, pp. 446.

3. Ashakina, A., Hasanuzzaman, M., Arifuzzaman, M., Rahman, M. W., & Kabir, M. L., (2016). Performance of single, double and three-way cross hybrids in tomato (*Lycopersicon esculentum* Mill.), *Food, Agriculture and Environment (JFAE)*, *14*(1), 71–77.

4. Vogel, K. E., (2009). Backcross breeding. *Methods Mol. Biol.*, *526*, 161–169. doi: 10.1007/978-1-59745-494-0-14.

5. Harrison, G. S., (1948). "What Is "Heterosis"?." *Genetics*, *33*(5), 439–446.

6. Schnable, P. S., & Nathan, M., (2013). Progress towards understanding heterosis in crop plants. Springer, *Annual review of Plant Biology*, *64*, 71–88.

7. Bisognin, D. A., (2011). Breeding vegetatively propagated horticultural crops. *Crop Breed. Appl. Biotechnol.*, *11*, no.spe Viçosa. http://dx.doi.org/10.1590/S1984–70332011000500006.

8. Zhao, J. H. X., Laroche, A., Lu, Z. H., Liu, H. K., & Li, Z., (2014). Genotyping-by-sequencing (GBS), an ultimate marker-assisted selection (MAS) tool to accelerate plant breeding. *Front. Plant Sci.*, *5*, 484. http://doi.org/10.3389/fpls.2014.00484.

9. Elshire, R. J., Glaubitz, J. C., Sun, Q., Poland, J. A., Kawamoto, K., Buckler, E. S., et al., (2011). A robust, simple genotyping-by-sequencing (GBS) approach for high diversity species. *PLoS One*, *6*(5), e19379. doi:10.1371/journal.pone.0019379 .

10. Cooper, M., Messina, C. D., Dean Podlich, D., Totir, L. R., Baumgarten, A., & Hausmann, N. J., (2014). Deanne wright and geoffrey graham. Predicting the future of plant breeding: complementing empirical evaluation with genetic prediction. *Crop and Pasture Science*, *65*(4), 311–336.

11. Guy, K. A., & Robert, M., (2013). Plums, apricots, and their crosses: Organic and low-spray production. *NCAT Agriculture Specialists Published May ©NCAT IP386.*

12. Sneep, J., (1958). The breeding of hybrid varieties and the production of hybrid seed in spinach. *Euphytica.*, *7*(2), 119–122.

13. Gowers, S., (2000). A comparison of methods for hybrid seed production using self-incompatibility in swedes (Brassica napus ssp. napobrassica). *Euphytica*, *113*(3), pp. 207–210.

14. Birchler, J. A., Yao, H., Chudalayandi, S., Vaiman, D., & Veitia, R. A., (2010). Heterosis. *The Plant Cell.*, *22*(7), 2105–2112.

15. Chen, J. Z., (2013). Genomic and epigenetic insights into the molecular bases of heterosis. *Nat. Rev. Genetics*, *14*, 471–482.

16. Cheng, S. H., Zhuang, J. Y., Fan, Y. Y., Du, J. H., & Cao, L. Y., (2007). Progress in research and development on hybrid rice: A super domesticate in China. *Ann. Bot. (Lond)*, *100*(5), 959–966.

17. Chang, Z., Chen, Z., Wang, N., Xie, G., Lu, J., Yan, W., & Zhou, J., (2016). Construction of a male sterility system for hybrid rice breeding and seed production using a nuclear male sterility gene. *Proc. Natl. Acad. Sci.*, *113*(49), 14145–14150.

18. Singh, R., Ram, L., & Srivastava, R. P., (2012). A journey of hybrids in maize: An overview, *Indian Research Journal of Extension Education Special Issue. 1*, 340–344.

19. Harrington, J. F., & Douglas, J. E., (1970). *Seed Storage and Packaging*. National seeds corp. in cooperation with the Rockefeller Foundation–Science, New Delhi, India, pp. 3–9.

20. Jyoti, N., & Malik, C. P., (2013). Seed deterioration: A review. *Int. J. Life Sc. Bt & Pharm. Res.*, *2*, 3.

21. Roberts, E. H. Storage environment and the control of viability. *Viability of Seeds*, pp. 14–58.

22. Rattan Lal Aggarwal, (2008). *Seed Technology*. Second edition. Oxford and IBH Publishing.

CHAPTER 9

PLANT TISSUE CULTURE

ANJALI PRIYADARSHINI and PRERNA PANDEY

CONTENTS

9.1 INTRODUCTION

Plant tissue culture, which is aseptic culture of cells, tissues, organs, and/ or their components under well-defined physical and chemical conditions in vitro, has emerged as a very important tool in both the basic and applied science along with its various commercial application. The beginning of this particular branch of science can be traced back to the ideas of the German scientist Haberlandt at the onset of the 20th century. The early studies done have paved the path for techniques such as root cultures, embryo cultures, and the first true callus and/or tissue cultures. The period between the 1940s and the 1960s was marked by the development of new techniques and the improvement of those that were already in use (Table 9.1). The availability of these techniques led to the application of tissue culture to five broad areas:

TABLE 9.1 Contributions of Various Scientists in the Field of Plant Tissue Culture during Infancy of this Branch of Science

Year	Scientist	contribution
1902	Haberlandt	First attempt of *in vitro* plant cell culture
1904	Hannig	Culture of embryogenic tissue of crucifers
1922	Robbins	*In vitro* root culture
1925	Laibach	Zygotic embryo culture in Linum
1934	White	Tomato plant root culture
1939	Gautheret, White and Nobecourt	Establishment of indefinite callus culture
1941	Braun	Crown Gall tissue culture
1945	Loo	Stem tip culture
1955	Miller	Discovery of kinetin hormone
1957	Skoog, Miller	Regulation of organ formation by Auxin:Cytokinin ratio discovery
1960	Bergmann	Development of plating technique for single cell isolation
1970	power	Protoplast fusion
1970	Guha and Maheshwari	Anther culture
1971	Takabe	Regeneration of plant from protoplast
1974	Reinhard	Biotransformation in plant tissue culture
1978	Melchers	Somatic hybrid Pomato production

1. Cell behavior that includes fields such as cytology, nutrition, metabolism, morphogenesis, embryogenesis, and pathology;
2. Plant modification and improvement;
3. Pathogen-free plants and germplasm storage;
4. Clonal propagation; and
5. Product (mainly secondary metabolite) formation.

The seed sown in the mid-1960s in the area of plant tissue culture saw continued growth and expansion in the application of the in vitro technologies to an increasing number of plant species in 1990s. Cell cultures emerged and have remain till date as an important tool in the study of basic areas of plant biology and biochemistry; they have attained a major significance in studies related to molecular biology as well as agricultural biotechnology. The historical development of in vitro technologies such as micropropagation, somatic embryogenesis, synthetic seed production, plant regeneration through callus-mediated shoot organogenesis, adventitious shoot regeneration, anther culture, tetraploid induction, and genetic transformation and their applications have been dealt with in the subsequent portion of this chapter.

Cultured plant cells are believed to generate genomic stress resulting from wounding, physicochemical factors, presence of hormones and/or enzymes, coupled with the developmental events of dedifferentiation and regeneration. These changes are manifested at the gene expression level in protein kinases, transcription factors (TF), and structural genes and contribute to explant adaptation to stress and reorientation of its developmental program. In vitro culture is also associated with genetic changes including chromosomal changes, DNA sequence alterations, amplifications, and transpositions. More recent discoveries point to epigenetic changes at the level of DNA methylation, chromatin modification, and small RNA-mediated regulation taking place in cultured tissues. The spectrum of genetic and epigenetic changes can potentially give rise to phenotypic changes among the regenerants that are termed somaclonal variations (Figure 9.1)

The past 30 years have witnessed a series of systematic biotechnological advances made in fruit crops such as pomegranate. Tobacco (*Nicotiana tabacum*) and its various related species have played a pioneer role in the development of plant biotechnology. Aseptic in vitro cultures were established as one of the first plant tissue in vitro cultures by the mid-20th century. In the turn of next decade, the tobacco cultures also served for investigations of

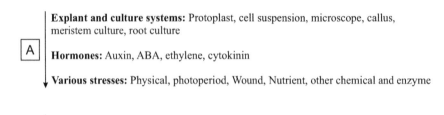

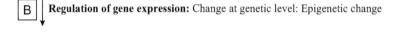

A-   In vitro environment

B-   Changes occurring at molecular level

C-   Application of cell fates

FIGURE 9.1   A flowchart of in vitro culture and molecular changes caused in the process [1].

organogenesis, which means formation of adventitious shoots and roots that are very important for micropropagation of plants. In the 1960s, the totipotency of plant cells, i.e., the ability for regeneration of the full plant from single cell has been unequivocally proved on the basis of experiments done with tobacco cell cultures in which entire plant was regenerated by tissue culture. Moreover, we have another reason to be thankful to tobacco in vitro cultures for the development of the medium for performing plant *in vitro* cultures, which was elaborated in 1962 and is still the most frequently used medium.

There are many firsts associated with tobacco as the experimental plant such as those listed below:

a) tobacco was the first plant regenerated from a protoplast.

b) the first somatic hybrid and chimera in vitro were derived from tobacco cultures (the 70s).

c) tobacco was also one of the first transgenic plants.

Thus, for the formation of transgenic plants, in vitro culture has been of utmost importance to carry out transformation of tissue and later regeneration of shoots and roots. It is thus a very logical assumption that preformed cell suspension culture of tobacco exhibited, on the one hand utility for studies of plant cell biology, and on the other hand, was applied for performing controlled biosynthesis of useful metabolites and biotransformation (bioconversion), that is directed chemical modification of foreign compounds introduced to such culture. It can be said that tobacco has played the role of the model plant in investigations of biology and biotechnological applications of plants.

The plant tissue culture methodology is a long process that many a times require great deal of precision as well as thorough inputs from technical as well as scientific experts. The general layout of the entire process is shown in Figure 9.2.

9.2 THE BASIC STEPS FOR IN VITRO CULTURING OF PLANTS

9.2.1 SELECTION AND STERILIZATION OF EXPLANT

A suitable explant is selected. Care has to be taken that this part is healthy and disease-free, which is then excised from the donor plant. The explant is then sterilized using disinfectants.

9.2.2 PREPARATION AND STERILIZATION OF CULTURE MEDIUM

A suitable culture medium is prepared depending upon the objectives of culture and the type of explant to be cultured because each use and part of plant to be taken for culturing has specific nutritional and growth requirements. Prepared culture medium is transferred into sterilized vessels and then sterilized in an autoclave.

9.2.3 INOCULATION

The sterilized explant is inoculated (transferred) on the adequate culture medium aseptically in a sterile environment. A lot of care needs to be undertaken to make the environment sterile and germ-free. An undifferentiated mass of cells develops at this stage, which is termed as callus.

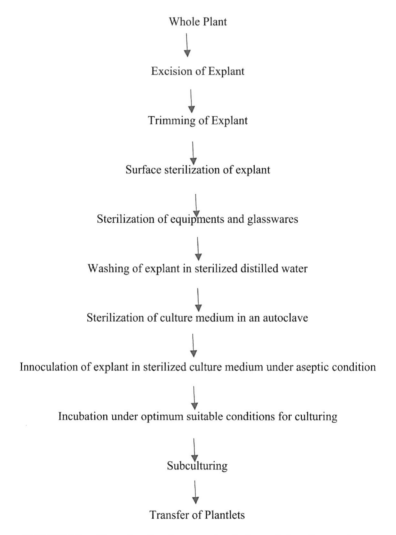

FIGURE 9.2   Chart showing the general technique of plant tissue culture.

9.2.4   INCUBATION

Cultures are then incubated in the culture room. The culture room/area has adequate light, temperature, and humidity conditions for successful culturing. This leads to the development of primary culture to be used in the following steps.

9.2.4.1   Subculturing

The cultured cells are then transferred to a fresh nutrient medium where the plantlets are obtained to be grown in green house.

9.2.4.2   Transfer of Plantlets

The plantlets are transferred to green house or in pots after the hardening process (i.e., acclimatization of plantlet to the environment).

9.3   CELLULAR TOTIPOTENCY

The term cellular totipotency is always used in case of plant cell; it can be defined as "the potential of a plant cell to grow and develop into a whole new multicellular plant." This phenomenon of totipotency can be explained in other words as the ability of a single cell to differentiate into many other cell types. This is the property that is exclusive to living plant cells and not animal cells (except stem cells in animals). The term totipotency was coined by Morgan in 1901. During culture practice, an explant is taken from a differentiated, mature tissue. This suffices the totipotency of a plant cell where a terminally differentiated non-dividing and quiescent cell reverts to its dividing undifferentiated form under adequate growth conditions (Figure 9.3).

To express its totipotency, as already explained, differentiated cell first must undergo the phenomenon of dedifferentiation and then through the re-differentiation phenomenon (Figure 9.3). Callus formation is the usual outcome of differentiation of explants. However, the embryonic explants, sometimes act as an exception to this rule where the differentiation of roots or shoots occurs without an intermediary callus state.

Thus, from the above account, it is clear that unlike animals (which shows irreversible differentiation except for stem cells), the plants have a particular property where highly mature and differentiated cells have an ability to revert to meristematic state and thus give rise to the entire structure. The property of totipotency of plant cells indicate that even the undifferentiated cells of a callus carry the essential genetic information required for regeneration of a whole plant.

It is also clear that all the genes responsible for dedifferentiation or redifferentiation are present within all the individual cells, which become

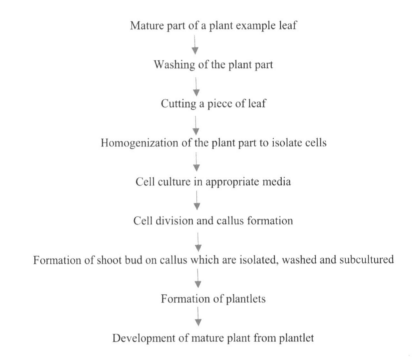

Mature part of a plant example leaf

Washing of the plant part

Cutting a piece of leaf

Homogenization of the plant part to isolate cells

Cell culture in appropriate media

Cell division and callus formation

Formation of shoot bud on callus which are isolated, washed and subcultured

Formation of plantlets

Development of mature plant from plantlet

FIGURE 9.3   Totipotency in plant cells.

active for expression under adequate culture conditions. As totipotent cells are the basis of whole plant tissue culture techniques, by the exploitation of this potential of plant cells, biotechnologists are trying to improve the crop plants and other commercially important plants.

9.3.1   *TOTIPOTENCY IN DIFFERENT PLANT PARTS*

Different plant parts have different totipotent abilities. For example, in tobacco plant, the type of buds formed by in vitro culture of the epidermis of different regions of the plant are different in their forms.

9.3.2   *APPLICATIONS OF TOTIPOTENCY*

Cellular totipotency of plants cells has proved to be a boon to mankind as it is the basis of plant tissue culture. The plant tissue culture exploits this unique property of plant cells to attain commercial benefits. Various applications of cellular totipotency are:

- Potential applications in the crop plant improvement
- Micro-propagation of commercially important plants.
- Production of artificial or synthetic seeds.
- Conservation of germplasm (formation of genetic resources).
- Production of haploid.
- Producing of somatic hybrids and cybrids.
- Helpful in cultivation of those plants whose seeds are very minute and difficult to germinate.
- Facilitates the study of cytological and histological differentiations.
- Can play a significant role for high scale and efficient production of secondary metabolites.
- Facilitates the possibility for genotypic modification.

9.4   DEDIFFERENTIATION

The phenomenon of reversal to meristematic dividing state from mature nondividing cells is called dedifferentiation. These dedifferentiated cells have the ability to form a whole plant or plant organ from cells or callus. This phenomenon is termed as re-differentiation. Thus, de-differentiation and re-differentiation are the two inherent phenomena that are involved in the property of cellular totipotency (Figure 9.4). It can be said that cell differentiation is the basic event for development of plants by tissue culture techniques, and it is also called cytodifferentiation.

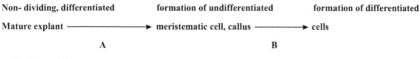

A: Dedifferentiation;

B: Redifferentiation

FIGURE   9.4   Schematic   representation   of   the   process   of   de-differentiation   and re-differentiation.

9.5   DIFFERENTIATION

The term differentiation describes the development of different cell types as well as the development of organized structures like roots, shoots, buds, etc., from cultured cells or tissue. Differentiation may also be defined in simple words as the development change of a cell that leads to its performance of specialized function. For example, differentiation accounts for the origin of different types of cells, tissues, and organs during the formation of a complete multicellular organism (or an organ) from a single-celled zygote.

Actually, the development of an adult organism starting from a single cell occurs as a result of the combined functioning of cell division and cell differentiation. Various techniques of tissue culture provide not only a scope of studying the factors governing totipotency of cells but also serves for the investigation of patterns and factors controlling the differentiation.

9.5.1   TYPES OF DIFFERENTIATION

As stated earlier also, the plant cells have a tendency to remain in a quiescent stage, which may be reverted to the meristematic stage. This process is termed as dedifferentiation, and as a result of this, a homogeneous undifferentiated mass of tissue, i.e., callus is formed. The callus cells then differentiate into different types of cells or an organ or an embryo.

On this basis, the differentiation may be of the following types:

- cytodifferentiation;
- organ differentiation;
- embryogenic differentiation.

9.5.1.1   Cytodifferentiation

Development is the outcome of cellular differentiation. Cytodifferentiation is the differentiation of different types of cells from the cultured cells. Callus which is an undifferentiated mass undergoes cytodifferentiation and re-differentiates into whole plant.

Amongst different cytodifferentiations, the differentiation into vascular tissues has received maximum attention. However, it is important here to mention that the cells of mature xylem elements and phloem cells cannot be re-differentiated or cannot be reverted to the meristematic state due to lack of cytoplasm in them.

Although in initial stages of their development, they can be reverted to meristematic cells. Xylogenesis is the differentiation of parenchymatous cells (of callus) into xylem-like cells of vascular plants. Phloem differentiation is the formation of phloem-cells from parenchyma in culture.

The factors affecting cytodifferentiation are:

i.   *Physical factors* like light, temperature, and pH are effective at optimum levels.
ii. *Chemical factors* like:
     a.   Low nitrogen content increases vascularization.
     b.   High Ca^{++} ions stimulates the formation of tracheid's and sieve tubes.
     c.   Sucrose in high concentration results in pronounced xylem differentiation.
iii. *Hormones*: Some hormones play an important role in cytodifferentiation. These are:
     a.   Auxin plays a major role in vascularization.
     b.   Cytokinin promotes cytodifferentiation.
     c.   Gibberellins along with auxins promote it.
     d.   Abscisic acid inhibits it usually.

Abscisic acid (ABA) plays a significant role in the regulation of many physiological processes of plants. It is often used in tissue culture systems to promote somatic embryogenesis and enhance somatic embryo quality by increasing desiccation tolerance and preventing precocious germination. ABA is also employed to induce somatic embryos to enter a quiescent state in plant tissue culture systems and during synthetic seed research. Application of exogenous ABA improves in vitro conservation and the adaptive response of plant cell and tissues to various environmental stresses. ABA can act as anti-transpirant during the acclimatization of tissue culture-raised plantlets and reduces relative water loss of leaves during the ex vitro transfer of plantlets even when non-functional stomata are present [2].

9.5.1.2   Organ Differentiation

It is synonymous to organogenesis or organogenic differentiation. It refers to the development or regeneration of a complete organized structure (or whole plant) from the cultured cells/tissues (Figure 9.5).

Organogenesis literally means the birth of organ or the formation of organ. It may occur either by shoot bud differentiation or by the formation of root. Organogenesis commences with the stimulus produced by the components of culture medium, the substances initially present in the original explants, and by the compounds produced during culturing. Roots, shoots, flower buds and leaves are among different organs that can be induced in plant tissue culture. Regenerations into flower buds and leaves occur in a very low frequency. However, the roots and shoot bud regenerations are quite frequent. Out of all these types of organogenic differentiation, only the shoot bud differentiation can give rise to the complete plantlets; therefore, it is of great importance in tissue culture practices.

The initiation of roots is termed as rhizogenesis, while the initiation of shoots is called as caulogenesis, and these two phenomena are affected by alterations in the auxin:cytokinin ratio in the nutrient medium. A group of

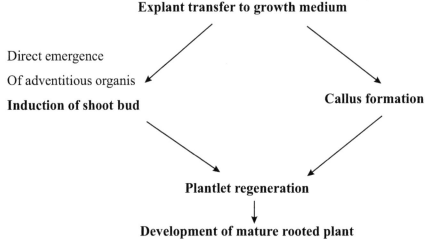

Explant transfer to growth medium

Direct emergence

Of adventitious organis

Induction of shoot bud

Callus formation

Plantlet regeneration

Development of mature rooted plant

FIGURE 9.5   The process of organogenesis in plant tissue culture.

meristematic cells called as meristemoids is the site of organogenesis in callus. Such meristemoids are capable of producing either a root or a shoot.

Factors affecting organogenesis:

- Auxin:cytokinin ratio in the medium is an important factor affecting root/shoot bud differentiation in most plants.
- Usually gibberellic acid inhibits organogenesis.
- Physiological state and size of explant play an important role in organ differentiation.
- Genotype of the donor plant plays a crucial role.
- Physical factors like light, temperature, moisture, etc., play an effective role in organogenesis.

9.5.1.3   Embryogenic Differentiation

The embryos formed from the somatic cells of plant in culture under in vitro conditions are called as somatic embryos. When the somatic cells of plant organs result into the regeneration into embryos, then the process is called as somatic embryogenesis or embryogenic differentiation or embryogenesis.

Somatic embryos are also referred to as embryoids, and they can be obtained either indirectly (with formation of callus) or directly from the explant without intervening callus formation. However, direct embryogenesis is not a normal process because the medium requirement for this is complex. Somatic embryogenesis is not used very frequently for the propagation of plants because the technique is usually difficult and there is a high risk of occurrence of mutations. Another major drawback of somatic embryogenesis is that there are greater chances of loss of regenerative capacity on repeated sub-culturing.

9.6   METHODS OF CULTURING PLANT MATERIAL

There are different methods of culturing plant material. These methods differ on the basis of explants used and their resultant products.

9.6.1   CELL CULTURE

Cell culture is the process of producing clones of a single cell. The clones of cell are the cells that have been derived from the single cell through mitosis and are identical to each other as well as to parental cell. First attempts for cell culture were made by Haberlandt in 1902 [3]. However, he failed to culture single cell, but his attempts stimulated other workers to achieve success in this direction.

The method of cell culture is meritorious over other methods of culturing because it serves as the best way to analyze and understand the cell metabolism and effects of different chemical substances on the cellular responses. Single cell culturing is of immense help in crop improvement programs through the extension of genetic engineering techniques in higher plants.

The method of cell culture is done by following three main steps:

i.   Isolation of single cell from the intact plant by using some enzymatic or mechanical methods.
ii. In vitro culturing of the single cell utilizing micro chamber technique, or micro drop method or Bergmann cell plating technique (Figure 9.6).
iii. Testing of cell viability done with the phase contrast microscopy or certain special dyes.

Cell suspension

↓

Isolation of single cell from cell aggregates by using fine gauze and filtering

↓

Addition of molten agar medium to the cells present in the filtrate

↓

Formation of colonies from single cell in the agar medium

FIGURE 9.6   Steps involved in cell culturing.

9.6.2   SUSPENSION CULTURE

A culture that consists of cells or cell aggregates initiated by placing callus tissues in an agitated liquid medium is called as a suspension culture. The continuous agitation of the liquid medium during a suspension culture is done by using a suitable device called as shaker, most common being the platform/orbital shaker. Agitation with shaker is important because it breaks the cell aggregates into single cell or smaller groups of cells, and it helps in maintaining the uniform distribution of single cell and groups of cells in the liquid medium. A good suspension is the one that has high proportion of single cells than the groups of cells. Changes in the nutritional composition of medium may also serve as a useful technique for breakage of larger cell clumps.

9.6.3   ROOT CULTURE

Pioneering attempts for root culture were made by Robbins and Kotte during 1920s [4]. Later, many workers tried for achieving successful root cultures. In 1934, it was White [5] who successfully cultured the continuously growing tomato root tips.

Subsequently, root culturing of a number of plant species of angiosperms as well as gymnosperms has been done successfully. Root cultures are usually not only helpful for giving rise to complete plants, but they also have importance's of their own. They provide beneficial information regarding the nutritional needs, physiological activities, nodulations, infections by different pathogenic bacteria or other microbes, etc.

9.6.4   SHOOT CULTURE

Shoot cultures have great applicability in the fields of horticulture, agriculture, and forestry. The practical application of this method was proposed by Morel [5] (1958) after they successfully recovered the complete Dahalia plant from shoot-tips cultures. Later, Morel realized that the technique of shoot culturing can prove to be a potent method for rapid propagation of plants (i.e. micro propagation). In this technique, the shoot apical meristem is cultured on a suitable nutrient medium. This is also referred to as meristem culture (Figure 9.7). The apical meristem of a shoot is the portion which is lying beyond the youngest leaf primordium. Meristem tip culture is also beneficial for recovery of pathogen-free specially virus-free plants through the tissue culture techniques. Various stages in this culture process are the initiation of culture, shoot multiplication, rooting of shoots and finally the transfer of plantlets to the pots or fields.

9.6.5 PROTOPLAST CULTURE

A protoplast is described as a plasma membrane bound vesicle that consists of a naked cell formed as a result of removal of cell wall. The cell wall can be removed by mechanical or enzymatic methods. In vitro culturing of protoplasts has immense applications in the field of plant biotechnology. It not only serves for genetic manipulations in plants but also for biochemical and metabolic studies in plants. For protoplast culture, firstly the protoplasts are isolated from the plants utilizing some chemical or enzymatic procedure. At present, there are available a number of enzymes which have enabled the isolation of protoplasts from almost every plant

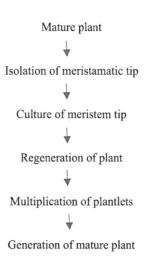

FIGURE 9.7   A general outlay of meristem culture.

tissue. After isolation of protoplasts, they are purified and then tested for their viability. Finally, the purified viable protoplasts are cultured in-vitro using suitable nutrient medium, which is usually either a liquid medium or a semisolid agar medium.

9.7 HAPLOID PRODUCTION

Haploid plants are those that contain half the number of chromosomes (denoted by n). Haploids can be exploited for benefits in the studies related to experimental embryogenesis, cytogenetics and plant breeding. Haploids have great significance in field of plant breeding and genetics. They are most useful as the source of homozygous lines.

In addition, the in vitro production of haploids also aids for induction of genetic variabilities, disease resistance, salt tolerance, insect resistance, etc. Presently, attention is being focused on improving the frequencies of haploid production in their advantageous utilization for economic plant improvement. There are two approaches for in vitro haploid production.

9.7.1 ANDROGENESIS

The technique of production of haploids through anther or microspore culture is termed as androgenesis. It is a method for the large-scale production of haploids through tissue culture. Androgenesis technique for haploid production requires the in vitro culture of male gametophyte, which is the microspore of a plant resulting into the production of complete plant from it. It is achieved either by anther culture or by microspore (pollen) culture (Figure 9.8).

9.7.2 GYNOGENESIS

It is an alternative source of in vitro haploid production. It refers to the production of haploid plant from ovary culture or ovule culture. The method of gynogenesis for haploid production has been successful, so far in a very few plants only; hence, it is not a very popular method for in vitro production of haploids. Thus, androgenesis is preferred over gynogenesis.

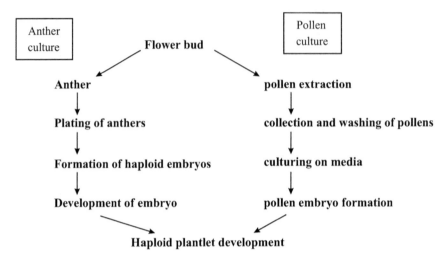

FIGURE 9.8   Haploid production by anther or pollen culture.

9.8   EMBRYO CULTURE

The technique of embryo culture involves the isolation and growth of an embryo under in vitro conditions to obtain a complete viable plant. Embryo culture is used widely in the fields of agriculture, horticulture, and forestry for the production of hybrid plants. This technique allows the detailed study about the nutritional requirements of embryos during different developmental stages. It also helps for identifying the regeneration potential of embryos. Embryo culture is advantageous for in vitro micro propagation of plants, overcoming seed dormancy, and for production of beneficial haploid plants.

9.9   ENDOSPERM CULTURE (TRIPLOID PRODUCTION)

Endosperm tissue is triploid; therefore, the plantlets originating by the culture of endosperm are also triploid. In majority of flowering plant families (exceptions being Orchidaceae, Podostemaceae, and Trapaceae, which lack endosperm), the endosperm tissues are present. Endosperm is formed after the double fertilization of one male nucleus with two polar nuclei.

Immature endosperm has more potential of growth in culture especially among the cereals.

Endosperm culture has provided a novel strategy for plant breeding and horticulture for the production of triploid plantlets. It is an easy method for production of a large number of triploids in one step.

Moreover, it is much more convenient that the conventional techniques like chromosome doubling by crossing tetraploids with diploids for triploid induction. Full triploid plants of endosperm origin have been produced in a number of plant species like *Populus*, *Oryza sativa*, *Emblica officinalis, Pyrus malus, Prunus,* etc.

The triploid plants are usually seedless; therefore, this technique is most beneficial for increasing the commercial value of fruits like apple, mango, grapes, watermelon, etc. In addition to all the above described applications, endosperm culture is also helpful for studying biosynthesis and metabolism of certain natural products.

9.10 SUBCULTURE OR SECONDARY CELL CULTURE

The process of transferring the cultured cells in a fresh nutrient medium is called as sub-culturing, and the cell cultures that are sub-cultured (i.e., inoculated in a separate medium) are called as subculture or secondary cell cultures or secondary cell lines.

It is important to subculture the organ and tissues to fresh medium to avoid the condition of nutrition depletion and drying of medium. It is possible to maintain the plant cell and tissue cultures for indefinitely long-time durations if they are regularly subculture in a serial manner.

9.11 SOMACLONAL VARIATION

Somaclonal variations are defined as the variations occurring in the cultured cells/tissues or plants regenerated from such cells in vitro. Somaclonal variations are usually heritable for qualitative as well as quantitative characteristics of plants. Somaclonal variants are being used as an alternate tool to plant breeding for the production of improved varieties of plants. Somaclonal variations are the result of gene mutations and changes in the structure and number of chromosomes. A number of new varieties of cereals, oil seeds, fruits, tomatoes, etc., possessing disease resistance, better quality, better yield, etc., have been generated through somaclonal variations. Some of those crop species are potato, tomato, oats, wheat, rice, maize, datura, carrot, soybean, etc.

9.12 SOMATIC HYBRIDS AND CYBRIDS

Somatic hybridization is described as the production of hybrid cells resulting by the fusion of protoplasts of somatic cells derived from two different plant species/varieties. It is very helpful in generating new and improved hybrid varieties of plant that may have characteristics of a completely different species as compared to the donor species.

For example, "Pomato" is a somatic hybrid that is produced by the fusion of protoplast of somatic cells from potato and tomato, which are totally different species. A cybrid is a cytoplasmically hybrid cell that has the cytoplasm of both fusing cells but the nucleus of only one fusing cell. The process of production of a cybrid is called cybridization.

Steps involved in somatic hybridization/cybridization (Figure 9.9) are given below:

Selection of the hybrid cells is done after fusion process. The selected somatic hybrids/cybrids are then verified for hybridity to check whether the hybrid is carrying the desired characteristics of both parents or not. After verification and characterization, the somatic hybrids or cybrids are cultured and regenerated into the plantlets with desired characteristics. The major advantage associated with somatic hybridization is that it overcomes the sexual incompatibility barriers and enables us to produce interspecific as well as inter-generic crosses in plants not occurring in nature. In addition, it also helps in providing disease resistance and improved quality characteristics in plants. Lastly, it has proved to be an immensely beneficial tool for the study of cytoplasmic genes and their expression.

9.13 MICRO-PROPAGATION

Tissue culture helps in the rapid propagation of plants by the technique of micro-propagation or clonal propagation in vitro. The asexually produced

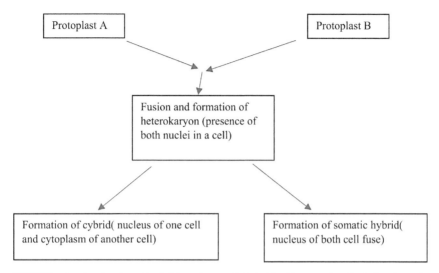

FIGURE 9.9   Production of cybrids and somatic hybrids. A) Protoplast isolation from the parent plant using any mechanical or enzymatic method. B) Fusion of isolated protoplasts derived from two different parents either by utilizing chemical fusogens (like NaNO$_3$, polyethylene glycol or high pH-high Ca^{++} treatment) or by the electro-fusion method.

progeny of a cell or individual is called as clone and the clones have an identical genotype.

Micro-propagation is the technique of in vitro production of the clones of plants, i.e., it produces the progeny plants that have an identical genotype as their parents, by cell, tissue or organ culture (Figure 9.10). It helps in the production of plants in large numbers starting from a single individual. It serves as an alternate method to conventional vegetative propagation methods.

Micro-propagation may be achieved in shoot tips, axillary buds, adventitious buds, bulbs, or somatic embryos. Some of the important plants that have been micro-propagated on large scales are:

- Orchids like Cymbidium, Dendrobium, Aranda, Vanda, Odontoma, Vanilla, etc.
- Forest trees like Tectona grandis, Biota, Cedrus deodara, Eucalyptus, Picea, Pinus, etc.

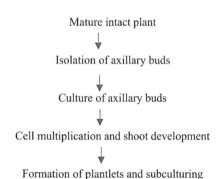

Mature intact plant

Isolation of axillary buds

Culture of axillary buds

Cell multiplication and shoot development

Formation of plantlets and subculturing

FIGURE 9.10   Micro-propagation: schematic representation.

The technique of micro-propagation generally involves four stages. Each of these stages has its own requirements. The stages in general technique of micro-propagation are described below:

- **Stage I** – Initiation: This stage also involves the preparatory process for achieving better establishment of aseptic cultures of explant. Suitable explant is selected from the mother plant. Then, the explant is sterilized and transferred to the nutrient medium for culture.
- **State II** – Multiplication: This is the most important stage of micro-propagation. In this stage, the proliferation or multiplication of shoots (or embryoids) occurs from the explant on the medium. It occurs either by the formation of an intermediary callus or by induction of adventitious buds directly from the explant.
- **Stage III** – Sub-culturing: The shoots are transferred to the rooting medium (sub-cultured) to form roots. As a result, complete plantlets are obtained.
- **Stage IV** – Transplantation: In this stage, the regenerated plantlets are transferred out of culture. These are grown in pots followed by field trials.

The advantages of micro-propagation are:

- Rapid multiplication of disease-free plants.
- Rapid multiplication of commercially important plants.
- Maintenance of genetic uniformity.
- Technique does not depend on seasons and is capable of producing

plants throughout the year.
- Technique is valuable in cases where only limited explant is available.

The limitations of micro-propagation are:

- The technique is costly.
- It requires proper skill.
- Many tree crops, including gymnosperms, cannot be multiplied by clonal propagation.
- Clonal propagation in some cases may lead to the formation of off-types rather than clones, after many generations.
- If culture is contaminated, then the pathogen gets multiplied to very high levels and becomes difficult to handle.

9.14   ARTIFICIAL SEED

These are also called as synthetic seeds. These are living seed like structures that are capable of giving rise to plants when sown in the field. An artificial seed is made of a somatic embryo (S.E.) encapsulated with a protective layer of a gel that protects it from desiccation or microbial attack (Figure. 9.11).

9.15   CRYOPRESERVATION: PRESERVATION AT ULTRALOW TEMPERATURE

This technique is used mainly for long-term storage of germplasm and thus helps in conservation of nature also. Plant tissues and organs are

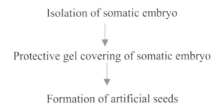

FIGURE 9.11   Schematic representation of steps involved in the preparation of artificial seed.

cryopreserved usually in liquid nitrogen, which has a temperature of −196°C. Cryopreservation technique has proved to be one of the most reliable methods for long-term storage and preservation of plant germplasm in the form of pollens, shoot-tips, embryos, callus, protoplasts, etc.

Although this is a very advantageous technique, it suffers from a major difficulty of formation of ice-crystals during freezing and/or thawing.

These ice-crystals may cause damage to the preserved material. To prevent the formation of ice-crystals during cryopreservation, some special chemicals called as cryoprotectants are used. A few common cryoprotectants are glycol, sucrose, proline, dimethyl sulfoxide (DMSO), polyethylene glycol (PEG), etc.

9.16 APPLICATIONS OF PLANT TISSUE CULTURE

- Germplasm conservation mainly in the form of cryopreservation of somatic embryos or shoot apices, etc.
- Large-scale production of useful compounds and secondary metabolites by using genetically engineered plant tissue cultures.
- Technique of micro-propagation for enhancing the rate of multiplication of economically important plants.
- Eradication of systemic diseases in plants and raising disease-free plants.
- Soma-clonal variations are useful sources of introduction of valuable genetic variations in plants.
- Helps plants in imparting resistance to antibiotics, drought, salinity, diseases, etc.
- Somatic hybrids and cybrids overcome species barriers and sexual incompatibility and produce hybrid plants with desired combination of traits.
- Embryo culture helps in overcoming seed sterility and dormancy.
- Haploid production in culture helps to solve various problems of genetic studies and thus aids the plant breeders for producing new varieties.
- Production of synthetic seeds via somatic embryo differentiation for commercially important plants helps to achieve increased agricultural production.
- Large-scale production of biomass energy.

- Plant tissue culture aids in producing the genetically transformed plants.
- Early flowering can be induced by in vitro culturing of plants so as to attain commercial benefits.
- Triploids as well as polyploid plants can also be produced by tissue culture techniques for use in plant breeding, horticulture, and forestry.
- Seedless fruits and vegetables can be produced by following the endosperm culture method, which adds to their commercial values.
- Increased nitrogen fixation ability can be achieved through association of tissue culture techniques with genetic engineering.
- Callus cultures are useful in plant pathology as they act as an effective tool in the study of mechanism of disease resistance and susceptibility.
- Different tissue culture techniques help us to study various biosynthetic processes, physiological changes, and cytogenetic changes.

9.17   MODIFICATION IN PLANT USING GENETIC ENGINEERING TOOLS

Rice is one of the world's most important food crops, and genetic modifications are extensively used for various purposes such as to increase yield and tolerate harsh environments. For several decades, rice has been modified by conventional breeding methods to produce plants with increased yields and greater resistance to pests and harsh weather conditions. Efforts are also being made to create rice plants with superior yield traits and resistance to biotic and abiotic stresses using genetic engineering techniques. Tissue culture has been heavily used for decades for transformation procedures to generate transgenic crops such as rice and maize [3].

Genetically modified plants are usually produced using tissue culture. New genes are introduced into plant cells that are growing in a dish, and each cell then replicates to form a mass of genetically identical cells. The application of plant hormones triggers the tissue to produce roots and shoots, giving rise to plantlet clones.

In addition to the genes that form its genome, the genetic make-up of an organism also includes its epigenome—a collection of chemical modifications that influence whether or not a given gene is expressed as a protein. The addition of methyl groups to specific sequences within the DNA, for example, acts as an epigenetic signal to reduce the transcription, and thus expression, of the genes concerned.

9.17.1   PLANT VECTORS

Scientists have many devices that allow them to achieve their goals of producing transgenic plants. In addition to various methods, such as the gene gun and biolistic techniques, natural bacterial vectors of plants such as *Agrobacterium tumefaciens* have been introduced as efficient tools, particularly in the case of cereals.

9.17.1.1   *Agrobacterium tumefaciens*

Agrobacterium tumefaciens is a soil plant pathogenic bacterium naturally infects the wound sites in dicotyledonous plant causing the formation of the crown gall tumors. It is widely known as the natural genetic engineer because this organism has the ability to transfer a portion of its DNA, also called transferred DNA (T-DNA), to the genome of a host plant [6]. Transformation of plant cell with a foreign gene was accomplished using *A. tumefaciens*. *A. tumefaciens*-mediated transformation has become the most used method for the introduction of foreign genes into plant cells and regeneration of transgenic plants with desired trait. Transformation of plant cell by *Agrobacterium* includes five essential steps as follows:

a) induction of the bacterial virulence system to initiate transfer;
b) generation of a T-DNA complex having the foreign DNA part to be transferred;
c) transfer of the T-DNA from *Agrobacterium* to the nucleus of the host cell for integration of foreign into the host genome;
d) integration of the T-DNA into the plant genome and generation of transformed cells;

e) the expression of T-DNA genes [7, 8] and selection of transgenics expressing the desired trait.

The *Agrobacterium* group constitutes virulent as well as nonvirulent strains. Strains of *A. tumefaciens and A. rhizogenes* are selected depending on their virulence to transform the plant cells. The virulent strains upon interacting with susceptible dicotyledonous plant cells, induce diseases known as crown gall and hairy roots, respectively, which is the visible effect showing that transformation has occurred. This property of virulent strains facilitates the selection of transformed part without the aid of complicated instruments or skilled labor.

The virulence of the strains is due to the presence of large megaplasmid (more than 200 kb) that play a key role in tumor induction. Due to this tumor inducing property, this megaplasmid was named Ti plasmid, or Ri in the case of *A. rhizogenes*. *A. tumefaciens* has the exceptional ability to transfer a particular DNA segment (T-DNA) of the tumor-inducing (Ti) plasmid into the nucleus of infected cells where it is integrated into the host genome and transcribed and cause crown gall disease [9]. T-DNA contains two types of genes:

a) First, the oncogenic genes encoding for enzymes involved in the synthesis of auxins and cytokinins and are responsible for tumor formation.

b) Second, the opine synthesis genes. These compounds (opines) are synthesized and excreted by the crown gall cells and consumed by *A. tumefaciens* as carbon and nitrogen sources, thus making the association beneficial for *Agrobacterium*. Ti plasmids are further classified according to the opines that are produced and excreted by the tumors they induce in dicotyledonous plants.

The T-DNA fragment is flanked by 25-bp direct repeats, which act as a *cis* element signal for the transfer apparatus; it is the most important part to transfer foreign DNA to the host genome (Figure 9.12). The process of T-DNA transfer is mediated by the cooperative action of proteins encoded by genes determined in the Ti plasmid virulence region (*vir* genes) in the bacterial chromosome. The 30-kb virulence (*vir*) region is a regulon organized in six operons that are essential for the T-DNA transfer (*virA, virB, virD,* and *virG)* or for the increasing of transfer efficiency (*virC* and *virE*) [10] (Figure 9.13).

The genes required for catabolism of opines are located outside the T-DNA. These are the genes involved in the process of T-DNA transfer

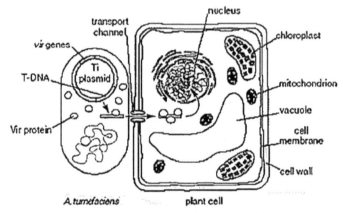

FIGURE 9.12   The process of transfer of gene in a plant cell by *Agrobacterium.*

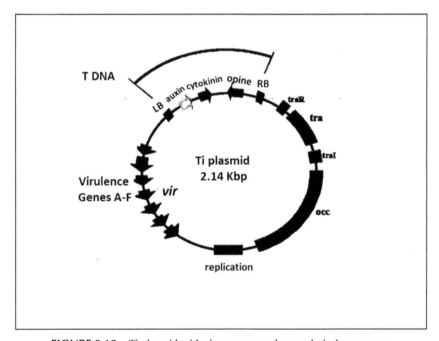

FIGURE 9.13   Ti plasmid with six operons and several virulence genes.

from the bacterium to the plant cell and the genes involved in bacterium-bacterium plasmid conjugative transfer [10].

There are three important facts that are used for the practical use of T-DNA transfer process for plants transformation for the development and construction of first vector and bacterial strain systems for plant transformation [11].

i.   Firstly, the tumor formation is a result of a transformation process due to transfer and integration of T-DNA into plant genome and the subsequent expression of T-DNA genes.
ii. Secondly, the T-DNA genes are transcribed only in plant cells and do not play any role during the transfer process.
iii. Thirdly, any foreign DNA placed between the T-DNA borders having repeat sequences can be transferred to plant cells, regardless of its origin.

The advantages of *Agrobacterium*-mediated transformation are:

i.   It reduces the copy number of the transgene, potentially leading to fewer problems with transgene co-suppression and instability (which is an outcome with development of many transgenic).
ii. Eliminates the possibility of development of mosaic plants otherwise obtained using traditional crop improvement methods where direct transformation is done. This is because it is a single-cell transformation system

However, *A. tumefaciens* naturally infects only dicotyledonous plants, and many economically important plants, including cereals, remained inaccessible for genetic manipulation for a long time. For these cases, alternative direct transformation methods have been developed [12], such as polyethylene glycol-mediated transfer, microinjection, protoplast and intact cell electroporation and gene gun technology.

Agrobacterium-based transformation methods have proved to be a very useful approach to genetically modify plants of various levels, which include plants such as the barrel clover (*Medicago truncatula*) *Arabidopsis* (*Arabidopsis thaliana*), and tobacco (*Nicotiana tobaccum*, *N. benthamiana*); cereals such as maize (*Zea mays*), rye (*Secale cereale*), barley (*Hordeum vulgare*), wheat (*Triticum aestivum*), rice (*Oryza sativa*), and sorghum (*Sorghum bicolor*); legume plants such as chickpea (*Cicer arietinum*); bean (*Phaseolus* spp.); pea (*Pisum sativum*); peanut (*Arachis hypogaea*), pigeon pea (*Cajanus cajan*); alfalfa (*Medicago sativa*), and many more.

9.17.2 TRANSGENIC SUGARCANE (SACCHARUM OFFICINARUM L.) MEDIATED BY AGROBACTERIUM

Sugarcane (*Saccharum officinarum* L.) is cultivated on large scale in tropical and subtropical regions as raw material for sugar and industrial products such as furfural, dextrans, and alcohol. Traditional plant breeding techniques, together with classic biotechnological approaches, have been extensively used to increase crop yields by selecting improved varieties that are more productive and resistant to diseases and pathogens. Some natural pharmaceutical compounds are derived from sugarcane [13]; additionally, agricultural and industrial by-products of the sugar production process are extensively employed for animal nutrition, food processing, paper manufacturing, and fuel [14]. Unfortunately, the genetic pool of sugarcane cultivars lacks some important traits such as resistance to insect pests and to some herbicides [15]. Thus, the use of plant transformation methods to introduce resistance genes into plant genomes, for example, development of first transgenic sugarcane lines resistant to stem-borer attack [15], can have an important impact on sugarcane yields. The lack of a reproducible methodology for stable transformation of sugarcane was an important obstacle for its genetic manipulation during many years. In 1992, Bower and Birch [16] successfully recovered transgenic sugarcane plants from cell suspensions and embryogenic calli transformed by particle bombardment. Later, a method to produce transgenic sugarcane plants by intact cell electroporation was established by Arencibia et al. [15].

The disadvantages of direct plant transformation systems are:

i.   traumatic to the cells;
ii. expensive due to the need of special equipment;
iii. poorly reproducible because of the variable transgene copy number per genome.

9.17.3   CONCERNS OF USING AGROBACTERIUM-BASED GENETIC TRANSFORMATION

The major concerns that have arisen regarding transgenic plants include the low efficacy of plant regeneration during tissue culture as well as production bottlenecks, such as in the spatial and temporal aspects of transgenic expression, target production, and the high level yield of recombinant products. Considering that *Agrobacterium* is a pathogen of dicotyledonous species, the efficacy of the *Agrobacterium*-based genetic transformation of monocotyledonous plants is still limited due to the low integration rate. The abovementioned difficulties prompted researchers to create new approaches to increase transformation efficiency. An *Agrobacterium* T-DNA derived nanocomplex has been introduced as a promising method for increasing the transformation efficiency of monocotyledonous plants [17]. This nanocomplex was first transferred into triticale (*Triticum × Secale*) microspores.

9.18   REVIEW QUESTIONS

1. Explain what plant tissue culture is and trace its development as an independent branch of science.
2. What are the various techniques to obtain a disease-free plant?
3. What is haploid culture. Elaborate.
4. What do you understand by somaclonal variation?
5. What is the role of *Agrobacterium tumefaciens* in genetic engineering of plant cell?
6. What is a Ti plasmid?

KEYWORDS

- artificial seed
- de-differentiation
- differentiation
- embryo culture

- **endosperm culture**
- **haploid production**
- **micropropagation**
- **somaclonal variation**
- **totipotency**

REFERENCES

1. Anjanasree, K., Neelakandan, A. K., et al., (2012). Recent progress in the understanding of tissue culture-induced genome level changes in plants and potential applications. *Plant Cell Rep.*, *31*, 597–620.

2. Manoj, K., Rai, N. S., Shekhawat, N. S., et al., (2011). The role of abscisic acid in plant tissue culture: a review of recent progress. *Plant Cell, Tissue and Organ Cult.*, *106*, 179–190.

3. Robbins, W. J., (1934). Cultivation of excised root tips and stem tips under sterile conditions. *Bot. Gaz.*, *73*, 376.

4. White, P. R., (1934). Potentially unlimited growth of excised tomato root tips in a liquid medium, *Plant Physiol.*, *9*, 585.

5. Morel, G., (1958). Biochemistry of morphogenesis of plant shoot, *Proc. 4th Int. Congr. Biochem.*, 221–222.

6. Sheng, J., & Citovsky, V., (*1996*). Agrobacterium-plant cell DNA transport: have virulence proteins, will travel. *Plant Cell.*, *8*(10), 1699–1710.

7. Gelvin, S. B., (2012). Traversing the cell: Agrobacterium T-DNA's journey to the host genome. *Front Plant Sci.*, *26*(3), 52.

8. Ziemienowicz, A., et al., (2012). A novel method of transgene delivery into triticale plants using the agrobacterium transferred DNA-derived nano-complex. *Plant Physiol.*, *158*(4), 1503–1513.

9. Nester, E. W., Gordon, M. P., et al., (1984). Crown gall: a molecular and physiological analysis. *Ann. Rev. Plant Physiol.*, *35*, 387–413.

10. Zupan, J. R., & Zambryski, P. C., (1995). Transfer of T-DNA from agrobacterium to the plant cell. *Plant Physiol.*, *107*, 1041–1047.

11. Torisky, R. S., Kovacs, L., et al., (1997). Development of a binary vector system for plant transformation based on super virulent *Agrobacterium tumefaciens* strain Chry5. *Plant Cell. Rep.*, *17*, 102–108.

12. Potrykus, I., (1991). Gene transfer to plants: Assessment of published approaches and results. *Ann. Rev. Plant Physiol. Plant Mol. Biol.*, *42*, 205–225.

13. Reiter, R. J., Tan, D. X., et al., (1994). Melatonin as a free radical scavenger: Implications for aging and age-related diseases. *Ann. New York Acad. Sci.*, *31*(719), 1–12.

14. Patrau, J. M., (1989). *By*-products of the cane sugar industry. An introduction to their industrial utilization. In: *Sugar Series,* Elsevier Science Publishers Amsterdam, B. V., Netherlands, *11*, pp. 435.

15. Arencibia, A., Roberto, V. I., et al., (1997). Transgenic sugarcane plants resistant to stem borer attack, *Mol. Breed.*, *3*(4), 247–255.

16. Robert, B., & Birch, B. G., (1992). Transgenic sugarcane plants via microprojectile bombardment. *Plant J.*, *2*, 409–416.

17. Ziemienowicz, A., et al., (2012). A novel method of transgene delivery into triticale plants using the agrobacterium transferred DNA-derived nano-complex. *Plant Physiol.*, *158*(4), 1503–1513.

CHAPTER 10

PLANT NUTRITION AND PHYTOHORMONES

PRERNA PANDEY and ANJALI PRIYADARSHINI

CONTENTS

10.1   INTRODUCTION

According to Allen Barker and Pilbeam [1], plants need macronutrients like N, K, P, Ca, S, and Mg and micronutrients like B, Cl, Mn, Fe, Zn, Cu, Mo, Ni, and Co for growth on medium (Table 10.1). Hence, the field of plant nutrition is of importance for agriculture to supply food to the world.

Plants achieve nutrient uptake from the soil via root hairs. The structure and architecture of the root can alter the rate of nutrient uptake. Nutrient ions are transported to the center of the root, the stele to facilitate nutrients to reach the conducting tissues, the xylem and phloem [3]. Carbon and oxygen are absorbed by the plants through the stomata.

An element is classified as "quasi-essential" when essentiality and plant responses vary among different plant species. Silicon is a quasi-essential element.

Plants uptake nutrients through mechanisms like simple diffusion, facilitated diffusion, and active transport. For certain nutrients, some plants enter into symbiotic relations with bacteria like *Rhizobia* for fixation of nitrogen or fungi to form mycorrhizae.

The field of plant nutrition is complex as there is variation among requirements for different species of plants. According to Huner and

TABLE 10.1   Essential Nutrients for the Plants [2]

Macronutrients	Micronutrients
Element (symbol): Ionic forms which are taken up by the plants	
Carbon (C): CO_2	Iron (Fe): Fe^{2+}, Fe^{3+}
Hydrogen (H): H_2O	Zinc (Zn): Zn^{2+}, $Zn(OH)_2$
Oxygen (O): O_2, H_2O	Manganese (Mn): Mn^{2+}
Nitrogen (N): NO_3^-, NH_4^+	Copper (Cu): Cu^{2+}
Sulfur (S): SO_4^{2-}	Boron (B): $B(OH)_3$
Potassium (K): K+	Molybdenum (Mo): $MoO4^{2-}$
Phosphorus (P): HPO_4^{2-}, $H_2PO_4^-$	Silicon (Si): $Si(OH)_4$
Magnesium (Mg): Mg^{2+}	Chlorine (Cl): Mg^{2+}
	Nickel (Ni): Ni^{2+}
	Cobalt (Co): Co^{2+}
	Vanadium (V): V^+

Hopkins [3], nutrients themselves show complex relations. For example, potassium ions uptake can be influenced by the amount of ammonium ions available. Let us briefly look at several nutrients and their roles in plants [4].

10.2 FORAGE LEGUMES: NUTRIENT SUPPLIERS

Among the nutrients required by plants, nitrogen has been found to be an important constituent of the cell. It is a constituent of amino acids that are the building blocks of proteins. Nitrogen has been established as an important constituent of chlorophyll [5].

Microorganisms may associate with several plant species and play their role in the nitrogen cycle. The N contribution from biological N2 fixation can reduce the need for industrial N fertilizers. Legumes generally do not require N fertilizer because of their symbiotic relationship with *Rhizobium* bacteria. Legumes and bacteria work together to extract atmospheric N (air contains 78% t N_2 but is unavailable to plants) and convert it to plant-available forms within legume roots [6]. Bacteria inside nodules convert free N to ammonia (NH_3), which the host plant utilizes to make amino acids and proteins.

10.2.1 FORAGE

According to Fageria [6] forage refers to crops that are eaten by livestock. Certain grasses used include *Agrostis* spp. – bentgrasses, *Arrhenatherum elatius* – false oat-grass, *Chloris gayana* – Rhodes grass, *Cynodon dactylon* – bermudagrass [7, 8].

Certain leguminous plants used as forage include *Medicago sativa* – alfalfa, Lucerne, *Trifolium* spp. – clovers, *Clitoria ternatea* – butterfly-pea, Vicia spp. – vetches, *Arachis pintoi* – pinto peanut to name a few [9, 10].

Rhizobia are bacteria capable of fixing N in association with a plant. Bacteria that nodulate legumes are currently classified into six genera (*Rhizobium, Bradyrhizobium, Mesorhizobium, Sinorhizobium, Azorhizobium,* and *Allorhizobium*) (Table 10.2).

TABLE 10.2   A Few Nitrogen-Fixing Bacteria and Their Hosts

Species	Host
Rhizobium leguminosarum	Broad beans, pea
R. ciceri	Chickpea
R. loti	Lotus
R. fredii, Bradyrhizobium japonicum, B. elkanii	Soyabean
R. meliloti	Alfalfa

10.2.2   NODULES

These are the tissue (small swellings) on the root system of legumes that house the rhizobia. Signals in the form of organic molecules, called flavonoids, from the plant communicate with the rhizobia. In most legumes, root hairs near the growing root tip curl as rhizobia attach to the tip, and an infection thread forms within the hair to allow bacterial entry. The plant produces a new meristematic region where rapid growth occurs to house the rhizobia and to provide them water and nutrients. The rhizobia, in return, use part of the nutrients provided to produce ammonia from nitrogen. The ammonia is then converted into an organic compound for transport and use by the plant [9, 10].

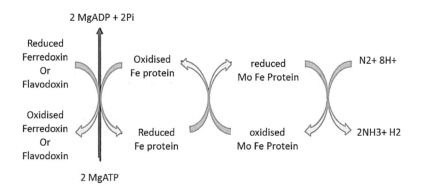

FIGURE 10.1   Mechanism of biological nitrogen fixation.

Research reports that nitrogen is the most limiting available nutrient in most agro-ecosystems, where mineralization and assimilation of N are crucial processes in plant–soil systems. Inclusion of legumes as a source of biological nitrogen fixation in agricultural systems supports in supplying N for non-legumes, while minimizing environment pollution associated with synthetic N-fertilizer application (Figure 10.1) [10].

10.3 MERITS AND DEMERITS OF FORAGE LEGUMES FOR RUMINANTS

Phelan et al. (2014) present merits and demerits of forage legumes for ruminants [11]. Merits include their high nitrogen and protein content, reduced dependence on fertilizers, and high intake and animal production. Few demerits such as higher bloat and difficulty in conservation as compared to silage has been reported. The primary advantage of forage legumes over other forages is their ability to reduce fertilizer N costs and their main disadvantage is usually lower intensity of animal production per ha of land when looked at from an economic aspect.

10.4 THERAPEUTIC VALUE OF FORAGE LEGUMES

Several phytochemicals including alkaloids and amines, cyanogenic glycosides, flavonoids, coumarins, condensed tannins, and saponins were investigated from several species like *Medicago sativa* (alfalfa), *Trifolium pratense* and *T. repens* (clovers), *Melilotus albus* and *M. officinalis* (sweet clovers), *Lotus corniculatus* (birdsfoot trefoil), *Onobrychis viciifolia* (sainfoin), *Lespedeza capitata* (roundhead lespedeza), and *Galega officinalis* (goat's rue). The therapeutic value of forage legumes has been broadly described. They report that several of these phytochemicals show antihypercholesterolemia, antidiabetic, antimenopause, anti-inflammatory, antiedema, anthelmintic, and kidney protective effects. Two widely prescribed drugs have been developed starting from temperate forage legumes, namely, the antithrombotic warfarin, inspired from sweet clover's coumarin, and the antidiabetic metformin, a derivative of sainfoin's guanidine. Further research is warranted in this field to explore the potential medicinal effects [12].

Chemicals known as condensed tannins from birdsfoot trefoil, BFT (*Lotus corniculatus* L.) and sainfoin (*Onobrychis viciifolia*), grouped in temperate forage legumes, as discussed by MacAdam et al. [13]. Their work demonstrated greater productivity of beef cattle fattened on BFT compared with other forages and greater milk production of commercial dairy cows grazing BFT compared with cows grazing grass pastures in mid-summer.

Agronomic benefits of rotating forage crops with annual grain crops include higher grain crop yields following forages. Perennial legumes in rotation also reduce energy requirements by adding significant amounts of N to the soil. Soil water availability may limit the extent to which forages benefit following crops. Under semiarid conditions, forages can actually reduce yields of the following crops, and as such, tillage practices that conserve soil water have been developed to partially address this problem. Forages in rotation provide environmental benefits, such as carbon sequestration, critical habitat for wildlife, and reduced NO_3 leaching [14].

Thilakarathna et al. examined the use of forage legume red clover cultivars to companion bluegrass and suggest the potential for developing red clover improving N-transfer to companion grasses while minimizing N losses through leaching [15].

10.5 ALLEY CROPPING: WHERE IT STANDS TODAY

Alley cropping is an agroforestry practice where agricultural or horticultural crops are grown in the alleyways between widely spaced rows of woody plant (USDA National Agroforestry Center First Edition, 2012) [16]. By combining annual and perennial crops that yield multiple products and profits at different times, a landowner can use available space, time and resources more effectively. For instance, the Midwest US include wheat, corn, soybeans or hay planted in between rows of black walnut or pecan trees.

Criteria for tree selection for alley cropping:
Annual or commercial product obtained periodically
Provide proper shade for the alley crop
Well-adapted to the site and soil
Foliage residue must not interfere with the alley crop
Growth requirements of the trees should complement the alley crops

FIGURE 10.2   A typical arrangement for alley cropping.

Alley cropping is also referred to as hedgerow intercropping as the trees that are managed as hedgerows, are grown in wide rows and the crop is planted in the interspace or "alley" between the tree rows (Figure 10.2).

One of the major benefits of alley cropping is the mulch provided by the hedgerow species, in the form of prunings, to the associated crop.

Many species have been used in the practice including *Leucaena leucocephala*, *Cajanus cajan*, *Flemingia macrophylla*, and *Erythrina peoppigiana* [17–19].

The USDA manual lists advantages of alley cropping such as an increase in a farmer's income, soil health improvement by certain species such as woody roots in the cropping system, and crops protected by the canopies of trees and shrubs [20].

Alley cropping has been shown to be advantageous in Africa, particularly in relation to improving maize yields in the sub-Saharan region. Use here relies upon the nitrogen- fixing tree species *Sesbania sesban*, *Euphorbia tricalii*, *Tephrosia vogelii*, *Gliricidia sepium*, and *Faidherbia albida*. In one example, a 10-year experiment in Malawi showed that, by using Gliricidia (*Gliricidia sepium*) without mineral fertilizer showed maize yields averaged 3.3 ton per hectare as compared to one ton per hectare in plots without fertilizer trees nor mineral fertilizers [21].

Gessesse et al. [22] reported that significantly higher grain and straw biomass yield of Teff (*EragrostisTeff*) were obtained through the application of *Sesbania sesban* and *Croton macrostaychus*. They suggested the employment of these two species and Teff in alley cropping systems [22].

In another report [23], finger millet, pigeonpea and peanut in between Leucaena supported mycorrhizal parameters like spore numbers and infective propagules of AM fungi in the rhizospheric soil and higher microbial C biomass grown between Gliricidia. They also reported better growing of finger millet when grown between Gliricidia [23].

10.6   GREEN MANURE: POTENTIALS

The dictionary defines the term as green manure is created by leaving uprooted or sown crop parts to wither on a field so that they serve as a mulch and soil amendment.

The Agriculture Information Bank defines it as, "a practice of ploughing in the green plant tissues grown in the field or adding green plants with tender twigs or leaves from outside and incorporating them into the soil for improving the physical structure as well as fertility of the soil." It can be defined as a practice of ploughing or turning into the soil, undecomposed green plant tissues for the purpose of improving the soil fertility.

As nitrogen is an essential element, the use of leguminous plants as green manure, the amount of nitrogen released into the soil lies between 40 and 200 pounds per acre. With green manure use, the amount of nitrogen that is available to the succeeding crop is usually in the range of 40–60% of the total amount of nitrogen that is contained within the green manure crop [24]. In India, ancient texts like Vrikshayurveda mention green manure.

Sustainable production of rice required introduction of green manuring in rice-based cropping system. Green manuring increases the nitrogen supply and improves the soil health. It also enhances the organic matter contents as well as increases the fertilizer use efficiency [25].

Both legumes and nonlegumes are used for making green manures. Examples include alfalfa, oats, rye, fenugreek, clover, millet, sunhemp, and Azolla [26]. The use of *Phacelia tanacetifolia* has also been reported [26].

Sesbania cannabina is the most widely used pre-rice green manure for rice in the humid tropics of Africa and Asia. *Astragalus sinicus* is the prototype post-rice green manure species for the cool tropics. Stem-nodulating *S. rostrata* has been most prominent in recent research. Many green manure legumes show a high N accumulation (80–100 kg N ha-1 in 45–60 days of growth) of which the major portion (about 80%) is derived from biological N_2 fixation. The average amounts of N accumulated by green manures can entirely substitute for mineral fertilizer N at current average application rates (reviewed by Becker [26]).

10.6.1 ADVANTAGES OF GREEN MANURE

The advantages include increased organic matter and soil nutrients, microbial activity from incorporation of cover crops into the soil leads to the formation of mycelium and viscous materials, which benefit the health of the soil by increasing its soil structure (from restoring the soil by brunch). The fixation of nitrogen by leguminous plants has its advantages. According to the Krishi Sewa, Sesbania can produce up to 80–100 kg N/ha (equivalent to 4–5 t dry biomass of Sesbania per ha) in around 40 days during the long-day season and in 50–60 days during the short-day season [26, 27].

There are reports of functions of weed suppression, for which non-leguminous crops (e.g., buckwheat) are primarily used. The deep rooting properties of many green manure crops make them efficient at suppressing weeds [28].

Islam et al. reported that the use of Sesbania green manure incorporated at 50 days after sowing in combination with 75% recommended dose of nitrogen could be considered more effective for BINA dhan7 rice production [29].

10.6.2 THE EFFECT OF GREEN MANURE SPECIES

Alfalfa and broad bean (*Vicia faba* L.) along with N fertilizers on rice paddy was studied by Gao et al. in 2015 [30]. The combined treatment of alfalfa and N fertilizer achieved the highest grain yield of all treatments. They suggested that green manure with a high P concentration is capable of alleviating P deficiency in both conventional tillage (CT) and no-till (NT) systems rice cultivation systems.

Hwang et al. [31] studied the effect of Legume hairy vetch and non-legume barley mixtures as green manure on rice cultivation and reported that the rice productivity increased for the mixed seeding of barley and hairy vetch at 25 and 75%. They suggested that this could be a useful agronomic practice for increasing the cover crop biomass and its nutrient productivity and to improve the productivity of subsequent rice crops in temperate rice paddies [31].

FIGURE 10.3   An example of a commercially sold biofertilizer

Long-term rice-rice-green manure rotation shaped the microbial community in the rice rhizosphere, as reported by Zhang et al. [32]; in particular, some beneficial bacteria such as *Acinetobacter* and *Pseudomonas* accumulated in the rhizosphere of green manure treatments.

10.7 BIOFERTILIZERS: A GROUND CHECK

According to Tilman [33], boosting of crop yields has for several decades been the arena of the chemical industry [33]. The green revolution, which arose during the 1940s and 1960s, included the development of nitrogen fertilizer derived from the Haber–Bosch process, phosphates, and various other nutrients and pesticides (Figure 10.3).

Conventional chemical fertilizers show certain hazards like pollution especially by eutrophication. Biofertilizer is a substance that contains living microorganisms which, when applied to seeds, plant surfaces, or soil, colonize the rhizosphere or the interior of the plant and promote growth by increasing the supply or availability of primary nutrients to the host plant [34].

Organic farming is mostly dependent on the natural microflora of the soil, which constitutes all kinds of useful bacteria and fungi including the arbuscular mycorrhiza fungi (AMF) called plant growth promoting rhizobacteria (PGPR) [34, 35].

When biofertilizers are applied as seed or soil inoculants, they multiply and participate in nutrient cycling and benefit crop productivity. The agriculturally useful microbial populations cover plant growth promoting rhizobacteria, N_2-fixing cyanobacteria, mycorrhiza, plant disease- suppressive beneficial bacteria, stress- tolerant endophytes, and biodegrading microbes [36].

Biofertilizers are a supplementary component to soil and crop management traditions, *viz.,* crop rotation, organic adjustments, tillage maintenance, recycling of crop residue, soil fertility renovation and the biocontrol of pathogens and insect pests, whose operation can significantly be useful in maintaining the sustainability of various crop productions [37].

Azotobacter, Azospirillum, Rhizobium, cyanobacteria, phosphorus and potassium- solubilizing microorganisms, and mycorrhizae are some of the

PGPRs that were found to increase in the soil under no tillage or minimum tillage treatment [38, 39].

It has been reported that in rice, the addition of *Azotobacter, Azospirillum* and *Rhizobium* promotes the physiology and improves the root morphology [40].

Besides playing role in nitrogen fixation, *Azotobacter* has the capacity to produce vitamins such as thiamine and riboflavin [41].

Plant hormones, *viz.,* indole acetic acid (IAA), gibberellins (GA), and cytokinins (CK) have also been produced by the organism [42].

Rhizobium has been used as an efficient nitrogen fixer for many years. It plays an important role in increasing yield by converting atmospheric nitrogen into usable forms [43].

Pseudomonas, Bacillus, Micrococcus, Flavobacterium, Fusarium, Sclerotium, Aspergillus, and Penicillium have been reported to be active in the solubilization process of phosphate and making it available to the plant [44]. *Enterobacter* and *Burkholderia* isolated from the rhizosphere of sunflower were found to produce siderophores and indolic compounds (ICs) which can solubilize phosphate [45].

Potassium- solubilizing microorganisms (KSM) such as *Aspergillus, Bacillus, and Clostridium* are found to be efficient in potassium solubilization in the soil and mobilize in different crops [46].

Apart from their role in nutrient fixation, an enhancement in resistance to stress, biotic and abiotic factors, to plants have been reported. *Rhizobium trifolii* inoculated with *Trifolium alexanderium s*howed higher biomass and increased number of nodulation under salinity stress condition [47].

Increased photosynthetic efficiency and the antioxidative response of rice plants subjected to drought stress were found after inoculation of arbuscular mycorrhiza [48].

Interestingly, it was reported that rhizoremediation of petroleum contaminated soil can be expedited by adding microbes in the form of effective microbial agent (EMA) to the few plant species such as cotton, ryegrass, tall fescue, and alfalfa [49].

By the production of certain metabolites, these organisms used as biofertilizers also provide resistance to pathogenic attack. *Bacillus subtilis*

GBO3 can induce defense-related pathways viz, salicylic acid (SA) and jasmonic acid (JA) [50].

In some cases, it was reported that along with bacteria, mycorrhizae can also confer resistant against fungal pathogens and inhibit the growth of many root pathogens such as *R. solani, Pythium spp., F. oxysporum, A. obscura and Heterobasidion annosum [51].*

The role of biofertilizers in regulating the activity of methane oxidizing bacteria, thereby moderating the methane sink in the soil have also been reviewed [52]. This study is of importance as the use of chemical fertilizers has resulted in increased emission of methane from the soil that can be prevented by such biofertilizers.

10.8   INTERACTIONS BETWEEN PLANT AND MICRO ORGANISMS

The major categories of plant–microbe interactions: the symbiosis between land plants and arbuscular mycorrhizae, nitrogen fixation by rhizobia within the nodules of legume roots and pathogenesis. The plant-microbe interaction is very complex with consortia of diverse families of microbes [53].

There is chemical signaling between plants and microbes during their interaction with action from both ends. It has been reported that there is rhizodeposition in the plants that modulates the bacterial species in the rhizosphere (reviewed by Farrar et al., 2014) [54].

Both root nodule-forming rhizobacteria and AM fungi release modified lipochitooligosaccharides called Nod factors and Myc factors, respectively, and their recognition by plants activates the SYM pathway that is required for successful symbiosis [55]. The amino acids and carbohydrates released by the root could serve as potential chemo attractants involved in interactions with microbes [56].

10.8.1   THE ROLE OF PHYTOHORMONES IN PLANT– MICROBE INTERACTIONS

Bottini et al. [57] reported the production of gibberellins in plants by certain bacteria [57]. The regulation of the plant hormone ethylene in experiments by Straub et al. showed alteration of phytohormone production upon microbial interaction [58]. They used Miscanthus seedlings with *Herbaspirillum frisingense* GSF30^T (reported as an endophyte of grass).

In addition, it was also observed that 1-aminocyclopropane-1-carboxylate (ACC) deaminase producing bacteria could regulate levels of ethylene. The ACC deaminase breaks down ACC that is a precursor to ethylene production.

The role of inducing cytokinin production by *Sinorhizobium meliloti* and *Mesorhizobium loti* that are symbiotic microbes has also been reported [60]. The role of cytokinin as active agents involved as regulators of modifying the plant root and fungal hyphae in the case of mycorrhizae has been reported by Barker et al. [61].

10.8.2 NGS TECHNOLOGIES FOR THE ANALYSIS OF PLANT-MICROBE INTERACTIONS

The use of sequencing technologies especially Next Generation Sequencing in studying these interactions has been summarized by Knief [62]. She has reviewed the usefulness of NGS technologies for the analysis of plant-microbes. Claudia et al. focused on the application of NGS methods in targeted metagenomic approach, and analysis of microbial communities by amplicon sequencing. Genomic analyses of individual microbial strains of whole microbial communities provide insight into potential of plant associated microorganisms, actual metabolic activities and regulatory mechanisms of the microbial cells under given conditions.

10.9 NODULATION

Bacteria belonging to the genera *Rhizobium, Mesorhizobium, Sinorhizobium, Bradyrhizobium,* and *Azorhizobium* (collectively referred to as rhizobia) grow in the soil as free-living organisms but can also live as nitrogen-fixing symbionts inside root nodule cells of legume plants (Figure 10.4).

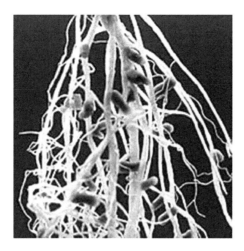

FIGURE 10.4   Roots of a leguminous plant showing nodules.

Symbiotic nitrogen- fixing bacteria are associated with certain plants like of the family Fabaceae in the "nodules." Apart from this family, the phenomenon has been reported in other families like actinorhizal plants [63].

Legumes form two different types of intracellular root symbioses, with fungi and bacteria, resulting in arbuscular mycorrhiza and nitrogen-fixing nodules, respectively.

Rhizobia carry most of the genes specifically required for nodulation either on large (500-kbp to 1.5-Mbp) plasmids. It is seen that nodules induced may be indeterminate (are elongated and give rise to new meristem that are repeatedly infected by the bacteria) or determinate (lack the repeated infection and meristem).

The role of flavonoids secreted by the plant in attracting the bacteria and expression of the *Nod* genes has been showed [64]. Other molecules such as the betaines (e.g., stachydrine and trigonelline) and the aldonic acids (e.g., erythronic acid and tetronic acid), are active as inducers in some rhizobial species at much higher concentrations.

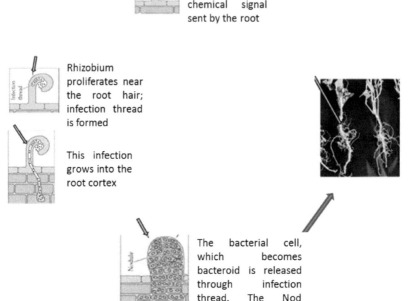

FIGURE 10.5   The process of rhizobial accumulation and nodulation in the host plant.

10.9.1   THE NOD GENES

The host root has been found to be associated with certain chemicals such as flavonoids that induce the expression of the Nod genes essential for synthesis of the Nod factor. Nod factor initiates many of the developmental changes seen in the host plant early in the nodulation process (Figure 10.5). Nod factors are lipo-chitooligosaccharides (LCOs). Nod factor synthesis depends on the expression of nodulation (*nod*) genes, comprising the *nod*, *nol*, and *noe* genes [65].

The NodC, NodB, and NodA proteins play a pivotal role in the synthesis of the LCO-backbone structure, by their function as chitin oligosaccharide synthase, chitin oligosaccharide deacetylase, and acyl transferase, respectively. NodD, which is a positive transcriptional regulator and mediates the identity of the flavonoid secreted by the plant. These nod factors are involved in the host –microbe interaction, infection, and formation of nodules.

NodI and NodJ are thought to be ABC- type transport proteins (reviewed by Spaink, 2000) [66]. *A nodF nodL* double mutant bacterium (these genes are involved in acetyl modification of the Nod factor) could not initiate infection threads in a host plant alfalfa [67].

The formation of mycorrhiza in plants also have shown genes called *DMI1*, *DMI2* and *DMI3* that control the expression of a nod factor pathway [67, 68].

10.9.2   THE PROCESS OF INVASION AND INFECTION OF RHIZOBIA [67, 68]

Nodule formation in roots of legumes begins when rhizobia attaches to the tip of a growing root-hair cell and secretes Nod factors (NF). These NF are sensed by the plant and series of events begins that include cytoskeletal rearrangements, curling at the root-hair tip, and formation of radially aligned cytoplasmic bridges called preinfection threads in outer cortical cells. An infection pocket filled with bacteria forms within the root-hair curl, from which the infection thread originates.

The infection thread extends down from the infection pocket into the root cortex, where it passes through pre-infection threads and eventually reaches the nascent nodule. The infection thread gets colonized by rhizobia, where they fix nitrogen. The Nod factors can also induce cortical root hairs. Several genes are involved in NF signaling and some of the downstream transcription factors required for infection have been characterized [68], some with roles in actin rearrangement and others with possible roles in protein turnover and secretion, including the roles of NF signaling, actin, and calcium and the influence of the hormones ethylene and cytokinin.

Azospirillum species are free-living N_2-fixing plant growth-promoting bacteria that live in close association with roots of cereals.

10.10   OXYGEN PARADOX

Most obligate or facultative aerobic N2-fixing bacteria have to deal with an "oxygen paradox." During oxygen –dependent growth, a minimal oxygen concentration is necessary to support aerobic respiration and ATP synthesis to meet the high-energy demand of nitrogenase. However, when present in too high concentration, oxygen can lead to irreversible damage of nitrogenase.

Nitrogenase is the enzyme complex responsible for nitrogen reduction; it is irreversibly inactivated by oxygen. Hence, this process requires conditions that are anoxic or nearly anoxic. In oxygen- rich environments, the nitrogenase is protected from inactivation by virtue of being located in differentiated cells with morphological and biochemical characteristics that reduce exposure of nitrogenase to oxygen. In some plants, root nodules develop to house nitrogen-fixing bacteria in a microaerobic environment. This process, a type of symbiotic nitrogen fixation is found with the genus *Rhizobia*.

Legume nodules resolve these paradoxical requirements of low free O_2 levels coupled with a rapid rate of O2 transfer to provide energy to the process by expressing high levels of leghemoglobins that are referred to as oxygen buffering proteins. Leghemoglobins accumulate to millimolar concentrations in the cytoplasm of infected plant cells prior to nitrogen fixation and are thought to buffer free oxygen in the nanomole range, avoiding inactivation of oxygen-labile nitrogenase while maintaining high oxygen flux for respiration. Ott et al. (2005) showed that leghemoglobins play a crucial role as buffers of oxygen in symbiotic Nitrogen fixation. The other mechanisms have also been reported as follows:

The bacteria reduce nitrogen to ammonia, but they themselves do not use the ammonia. This is linked to switching off of glutamine synthase that is involved in processing of ammonia by the plants that exert this control. This limits the growth of the bacteria. Another control is at the level of lack of the homocitrate gene. The bacteria would have to depend on the plant expressed product thereby are subject to control of their growth.

Certain legumes produce nodule-specific, cysteine-rich peptides (NCRs) that make the bacteria lose the ability to regenerate themselves but can efficiently fix nitrogen. The vitality of the bacteria would be thus linked to the mortality of the plant and would die if the host dies.

A mechanism called autoregulation of nodulation is exerted by legumes. A signal is translocated to shoots where it induces the production of a signal that is then translocated to roots to suppress the number of nodules formed.

Another level of control is at the level of nutrients as plants supply certain factors like malate, magnesium etc. to the bacteria [71].

10.11 STRATEGIES BY THE BACTERIA TO DEAL WITH OXYGEN PARADOX [72, 73]

1. The fixation of nitrogen has been coupled with the production of hydrogen whose high levels may inhibit the activity of nitrogenase. Diazotrophic bacteria show dehydrogenase activity to reduce this inhibition by hydrogen. *Streptomyces thermoautotrophicus* shows the presence of an oxygen tolerant nitrogenase system. Here, there is a link between reduction of nitrogen and oxidation of carbon monoxide and certain differences in the cofactors associated with nitrogenase [73].

2. Symbiotic Rhizobia with nodules form a diffusion barrier of nodules that restricts the O2 concentration in the infection zone.

3. Free- living bacteria that fix nitrogen have shown certain other mechanisms. There is a difference in regulation of the *nif* genes (genes responsible for N_2 fixation). A transcriptional or post translational control of NifA: a regulatory activator of the nif genes has been reported to alter the level of nitrogenase activity in response to oxygen levels. For instance, in *Azotobacter vinelandii* and *Azotobacter chroococcum*, a transient O2 increase leads to a reversible switching-off of nitrogenase [72–74].

10.12 PHYTOHORMONES: A REVIEW

The regulation of plant growth is controlled by phytohormones or plant growth regulators. They are present in algae and higher plants [75].

The hormones work as chemical messengers involved in communication of cellular activities in higher plants [76]. There are currently five primary classes of plant hormones: auxins, gibberellins, cytokinins, abscisic acid, and ethylene. Certain classes like the brassinosteroids, jasmonates, and strigolactone has also been discussed.

The various classes of hormones are summarized as in the following subsections.

10.12.1 AUXINS

Auxins were the first group to be discovered and despite extensive work their mode of action is still ambiguous [77]. IAA is produced mainly in the shoot apex bud and young leaves of plants. Additionally, flowers, fruits and young seeds have also been shown to be sites of hormone production. These hormones play a role in cell enlargement, bud formation, and root initiation. IAA (indole-3-acetic acid) is a vital phytohormone that is involved in growth regulation and also as a coordinator of plant growth under stress. It has been reported by Walz et al. that auxins in seeds regulate specific protein synthesis [78].

Auxins are thought to play a role in apical dominance, which is the inhibition of the development of some or all the lateral buds by the apical bud of shoot. Certain synthetic auxins (NAA: naphthalenacetic acid and indole butyric acid or IBA) find potential as rooting compounds [79].

Research indicates the role of IAA in co-ordination of plant growth during stress response to salinity. [80, 81] Briefly, at low auxin concentrations, auxin response factors (ARFs) that regulate auxin-responsive gene expression are repressed. By binding to the auxin-responsive element commonly found in the promoters of auxin-responsive genes, ARFs regulate (activate or repress) auxin-responsive gene expression, leading to a variety of auxin-mediated phenotypic alterations.

Moreover, microRNAs also regulate the expression of key genes involved in auxin signaling and lateral root development. More recently, IAR3 has been identified as a target of miR167 [82].

10.12.2 CYTOKININ

This group has a highly synergistic effect along with auxins: the ratios control the response of the plant. A key role of cytokinins has been seen in delay of senescence that involves the breakdown of macromolecules and chlorophyll. The induction of flowering and reversal of apical dominance has also been attributed to this class [83].

Cytokinins help in the release of seeds from dormancy [76, 80]. O'Brien and Benkova in 2013 have reported the roles of cytokinin along with Abscisic acid [83]. During water stress or drought, there is decrease in cytokinin and increase in Abscisic acid that controls the stomatal aperture.

Research indicates the role of cytokinins in stress response. The role of cytokinin to response of dehydration, drought, salt, and abscisic acid metabolism has been observed [84, 85].

10.12.3   GIBBERELLINS

They have been reported in fungi and plants and were discovered by researchers in a fungus called *Giberella fujikuroi*. Gibberellins are involved in a wide range of developmental responses. The group has been shown to break dormancy, delay senescence, and induce parthenocarpy. Another important role of gibberellins is facilitating endosperm mobilization through the induction of hydrolytic enzymes such as α-amylase and protease in the cereals and grass seeds. Other roles include sex determination, bolting of rosette plants, the promotion of seed germination, fruit development and the control of juvenility. In addition, gibberellins also have a positive role on seed germination, leaf expansion, stem elongation, flower and trichome initiation, and flower and fruit development [86].

10.12.4   ABSCISIC ACID (ABA)

This hormone finds its name as its role was recognized in fallen or abscised leaves. It has been reported as a stress hormone. Under water stress, ABA regulates the stomatal opening. Yan et al. [87] reported that reactive intermediates like nitric oxide is involved in ABA induction that mediates the sodium/potassium ion flux in guard cells affecting their turgidity [87]. This provides an application of ABA as an anti-transpirant in agriculture.

ABA also has a role to play in plants under stress. Its levels have been observed to be upregulated in plants under salinity stress [89]. ABA is involved in the regulation of expression of various stress related genes and protective proteins like late embryogenesis abundant (LEA) or dehydrins. LEA proteins are involved in protection from osmotic stress or desiccation and cold shock [90]. Dehydrins are stress proteins belonging to the LEA family.

ABA is involved in the production of dehydrins in response to several abiotic stresses [91].

10.12.5 ETHYLENE

It is a gaseous hydrocarbon: C2H4. It plays a key role in ripening of fruits. It is used in the storage of fruits after ripening-the treatment is given following a carbon dioxide rich treatment. It is now known that ethylene is produced by all living plant tissues and regulates their growth. The role of ethylene in the induction of epinasty is also observed during the flooding of plant roots, when leaf cells adopt a more vertical position than a horizontal position caused by the cells at the top to outgrow the ones at the bottom.

Low temperature and salinity have been shown to regulate the levels of ethylene thus showing its role as a plant stress regulator especially in heat stress [92, 93]. Depending on the species or the tissues involved, ethylene can have markedly different effects on development. For example, while the gas generally inhibits the growth of dicotyledonous shoots (e.g., peas), it promotes growth in a number of hydrophytes , e.g., rice. Equally, in the light, ethylene can actually promote the growth of *Arabidopsis* seedlings [92, 93].

10.12.6 BRASSINOSTEROIDS (BR)

These are polyhydroxysteroids that have been reported as plant hormones. Brassinosteroids, along with auxins play a role in cell elongation [94]. The roles played by these hormones are diverse including plant growth, seed germination, rhizogenesis, senescence, and resistance to plants against various abiotic stresses [95]. Caño-Delgado et al. [96] have studied the role of the steroids during cell differentiation in *Arabidopsis* [96]. Their role in vascular differentiation and signaling was examined. The role of brassinosteroids in the response of various stress factors have also been reported in several research works. BR has been observed to play a role in heat stress regulation in barley and in resistance to low temperature on rice, as well as in stress response through antioxidant response system [97].

10.12.7   NITRIC OXIDE

This signaling molecule has been involved in several defense systems like stomatal closure, root development, germination, nitrogen fixation, cell death, and stress response [98].

10.12.8   SALICYLIC ACID (SA)

Its role in various stress responses have been reported. SA mediates the phenylpropanoid pathway playing a role against pathogens and some insect pests, and abiotic stresses [99], in stress response such as drought, salinity, heat and chilling, and in regulation of water, salinity and low temperature stress [99].

10.12.9   JASMONATES

These are fatty acid derivatives that show a range of functions. They are involved in flowering, as well as cell death and leaf senescence which could limit the spread of infections. Jasmonic acid mediates the octadecanoid pathway meant for the defense against insect pests and some pathogens [100]. They have a role of the groups especially methyl jasmonate that elevates transcription of genes involved in wound response [101]. Wani, et al. [102] have reviewed the role of this group in various stress response like drought, salinity, low temperature, salinity, and heavy metal stress.

10.12.10   POLYAMINES

These are a group of polycations that have been implicated in stress response to metal toxicity, oxidative stress, drought, salinity, water logging, and low temperature stress [103], apart from their role in DNA replication and regulation of enzymes.

10.12.11   STRIGOLACTONES (SL)

These are a recently discovered group of phytohormones that play a role in root development, also in response to nitrogen and phosphorus. Their crosstalk with auxin during root development is also a potential sphere of research [104]. Foo and Davies [105] have reported that SL promote nodulation in pea, thus establishing them as important factors in nitrogen fixation [105].

10.12.12   TRIACONTANOL

It is a growth regulator found in epicuticular waxes of plants. Its application was reported to mediate increased uptake of nutrients, photosynthesis, and nitrogen fixation among several other benefits [106].

10.12.13   PLANT PEPTIDE HORMONES

Systemin is a small peptide that mediates protection of plants against herbivores. Systemin promotes the synthesis of antinutritional proteins, signaling pathway proteins and proteases [107], and in regulation of protein system to salt stress and UV radiation [108].

A peptide called stomagen is involved in stomata development, and overexpression or addition of stomagen has been shown to increase the number of stomata in plants [73].

A class of regulators called phytosulfonkines (PSKs) may serve to coordinate signals from interactions with pathogens and symbionts with the plant's growth requirement.

10.13   SUMMARY

Plant nutrition is a complex area of study as there is variation among requirements for different species of plants. There are several ways the plants improve nutrient availability and absorption. Nitrogen has been established as an important constituent of chlorophyll. Microorganisms may associate with several plant species and play their role in the nitrogen cycle, and biological N_2 fixation can reduce the need for industrial

N fertilizers. The primary advantage of forage legumes is their ability to reduce fertilizer N costs. Alley cropping is another practice where crops are grown in the alleyways between widely spaced rows of woody plant. Making green manure is a practice of ploughing the field for improving soil fertility. Symbiosis is one of the major categories of plant–microbe interactions involving land plants and nitrogen fixation by rhizobia within the nodules of legume roots and pathogenesis. The Nod proteins of leguminous plants play a pivotal role in plant physiology. Genomic analyses of individual microbial strains of whole microbial communities provide insight into potential of plant associated microorganisms, actual metabolic activities and regulatory mechanisms of the microbial cells under given conditions. The five primary classes and certain classes of plant hormones have also been discussed.

10.14 REVIEW QUESTIONS

1. Briefly describe the process of nitrogen fixation.
2. Describe a few therapeutic values of forage legumes.
3. What is alley cropping? How is it significant in today's times?
4. What are biofertilizers? How do they help in enhancing the soil quality?
5. What are *Nod* genes?
6. What is oxygen paradox and how do microbes deal with it?

KEYWORDS

- **biofertilizers**
- **green manure**
- **legumes**
- **oxygen paradox**
- **phytohormones**
- **plant nutrition**

REFERENCES

1. Allen, V. B., & Pilbeam, D. J., (2007). *Handbook of Plant Nutrition*. CRC Press. ISBN 978-0-8247-5904-9. Retrieved 17 August 2010.
2. Mengel, K., & Kirkby, E. A., (1982). *Principles of Plant Nutrition*, 3rd Edition. Published by International Potash Institute.
3. Norman, P. A. H., & William, H., (2008). *Introduction to Plant Physiology*, 4th edition. John Wiley & Sons, Inc.
4. Silva, J. A., & Uchida, R., (2000). Plant nutrient management in Hawaii's Soils, *Approaches for Tropical and Subtropical Agriculture*. Silva, J. A., & Uchida, R., eds. College of Tropical Agriculture and Human Resources, University of Hawaii at Manoa.
5. Roy, R. N., Finck, A., Blair, G. J., & Tandon, H. L. S., (2006). "Chapter 3: Plant nutrients and basics of plant nutrition." *Plant Nutrition for Food Security*: a guide for integrated nutrient management (PDF). Rome, Food and Agriculture Organization of the United Nations. pp. 25–42.
6. Fageria, N. K., (1997). *Growth and Mineral Nutrition of Field Crops*. New York, Marcel Dekker, pp. 595.
7. Murphy, B., (1998). *Greener Pastures on Your Side of the Fence*. Colchester, Vermont, Arriba Publishing, pp. 19–20.
8. Gilman, D. C., Thurston, H. T., & Colby, F. M., (1905). "*Pasture.*" New International Encyclopedia (1st ed.). New York, Dodd, Mead.
9. Beegle, D. B., Carton, D., & Bailey, J. S., (2000). Nutrient management planning: Justification, theory, practice. *Journal of Environmental Quality, 29*, 72–79.
10. Tom, L. *Nitrogen Fixation by Forage Legumes*, Iowa State University, Ames, IA. http://www.public.iastate.edu/~teloynac/354n2fix.pdf.
11. Phelan, P., Moloney, A. P., McGeough, E. J., Humphreys, J., Bertilsson, J., O'Riordan, E. G., & O'Kiely, P., (2015). Forage legumes for grazing and Conserving in ruminant production systems. *Critical Reviews in Plant Sciences, 34*(1–3), 281–326.
12. Laura, C., Jianbo, X., & Bruno, B., (2016). Therapeutic potential of temperate forage legumes: A review, *Critical Reviews in Food Science and Nutrition, 56*(1), 149–161.
13. Jennifer, W., MacAdam, & Juan, J., (2015). Villalba beneficial effects of temperate forage legumes that contain condensed tannins. *Agriculture, 5*(3), 475–491.
14. Martin, H. E., Vern, S. B., Patrick, M. C., Dwain, W. M., Ray, S. S., & McCaughey, P. W., (2002). Potential of forages to diversify cropping systems in the northern great plains, *Agronomy Journal, 94*(2), pp. 240–250.
15. Thilakarathna, M. S., Papadopoulos, Y. A., Rodd, A. V., Grimmett, M., Fillmore, S. A. E., Crouse, M., & Prithiviraj, B., (2016). Nitrogen fixation and transfer of red clover genotypes under legume–grass forage based production systems. *Nutrient Cycling in Agroecosystems, 106*(2), pp. 233–247.
16. What is Alley cropping? USDA National Agroforestry Center, (2012). First edition. February 2012. Available on: https://www.fs.usda.gov/nac/documents/workingtrees/infosheets/WT_Info_alley_cropping.pdf.
17. Kang, B. T., Reynolds, L., & Atta-Krah, A. N., (1990). Alley farming. *Advances in Agronomy, 43*, 315–359.

18. Kang, B. T., Gichuru, M., Hulugalle, N., & Swift, M. J., (1991). Soil constraints for sustainable upland crop production in humid and subhumid west Africa. In: *Soil Constraints on Sustainable Plant Production in the Tropics*, Tropical Agriculture Research Center, Tsukuba, Japan, pp. 101–112.

19. Kass, D. L., Araya, S. J. S., Sanchez, J. O., Pinto, L. S., & Ferreira, P., (1992). *Ten-Year Experience with Alley Farming in Central America*. Paper read at International Alley Farming Conference, IITA, Ibadan, Nigeria. Available on: https://slideheaven. com/alley-cropping-past-achievements-and-future-directions.html.

20. Kang, B. T., & Gutteridge, R. C. www.fao.org. Forage Tree Legumes in Alley Cropping Systems.

21. Akinnifesi, F. K., Makumba, W., & Kwesiga, F. R., (2006). Sustainable maize production using gliricidia/maize intercropping in southern Malawi. *Experimental Agriculture, 42*(4), 1–17.

22. Abrham, T. G., & Hailie, S. W., (2014). On-farm evaluation of multipurpose tree/ shrub species for sustaining productivity in alley cropping. *Malaysian Journal of Medical and Biological Research, 1*(3).

23. Balakrishna, A. N., Lakshmipathy, R., Bagyaraj, D. J., et al., (2016). Influence of alley copping system on AM fungi, microbial biomass C and yield of finger millet, peanut and pigeon pea. *Agroforest Syst.*, pp. 1–7.

24. Sullivan, P., (2003). Overview of cover crops and green manures, *Fundamentals of Sustainable Agriculture*. Available on: https://cpb-us-east-1-juc1ugur1qwqqqo4. stackpathdns.com/blogs.cornell.edu/dist/e/4211/files/2014/04/Overview-of-Cover-Crops-and-Green-Manures-19wvmad.pdf.

25. Shahzad, T., & Muhammad, I., (2013). *Green Manuring for Rice, 5*(9).

26. Becker, M., Ladha, J. K., & Ali, M., (1995). Plant Soil: Green manure technology: Potential, usage, and limitations. *A Case Study for Lowland Rice, 174*, 181.

27. *Restoring the Soil*, (2012). A guide for using green manure/cover crops to improve the food security of smallholder farmers by Roland Bunch. Canadian Foodgrains Bank Incorporated.

28. Vasilakoglou, I., Dhima, K., Anastassopoulos, E., Lithourgidis, A., Gougoulias, N., & Chouliaras, N., (2011). Oregano green manure for weed suppression in sustainable cotton and corn fields. *Weed Biology and Management, 11*, 38–48.

29. Islam, M. R., Hossain, M. B., Siddique, A. B., Rahman, M. T., & Malika, M., (2014). Contribution of green manure incorporation in combination with nitrogen fertilizer in rice production. *SAARC J. Agri., 12*(2), 134–142.

30. Gao, X., Lv, A., Zhou, P., Qian, Y., & An, Y., (2015). Effect of green manures on rice growth and plant nutrients under conventional and no-till systems. *Agron. J., 107*, 2335–2346. doi:10.2134/agronj15.0225.

31. Hyun, Y. H., Gil, W. K., Yong, B. L., Pil, J. K., Sang, Y. K., (2015). Improvement of the value of green manure via mixed hairy vetch and barley cultivation in temperate paddy soil. *Field Crops Research, 183*, pp. 138–146.

32. Xiaoxia, Z., Ruijie, Z., Jusheng, G., Xiucheng, W., Fenliang, F., Xiaotong, M., Huaqun, Y., Caiwen, Z., Kai, F., & Ye, Deng, (2017). Thirty-one years of rice-rice-green manure rotations shape the rhizosphere microbial community and enrich beneficial bacteria. *Soil Biology and Biochemistry, 104*, pp. 208–217.

33. Tilman, D., (1998). The greening of the green revolution. *Nature, 396*, 211–212.

34. Vessey, J. K., (2003). Plant growth promoting rhizobacteria as bio-fertilizers. *Plant Soil, 255*, 571–586.
35. Deepak, B., Mohammad, W. A., Ranjan, K. S., & Narendra, T., (2014). Microbial biofertilizers function as key player in sustainable agriculture by improving soil fertility, plant tolerance and crop productivity. *Cell Factories, 13*, 66. DOI: 10.1186/1475-2859-13-66.
36. Singh, J. S., Pandey, V. C., & Singh, D. P., (2011). Efficient soil microorganisms: a new dimension for sustainable agriculture andenvironmental development. *Agric. Ecosyst. Environ., 140*, 339–353. 10.1016/j.agee.2011.01.017.
37. Sahoo, R. K., Bhardwaj, D., & Tuteja, N., (2013b). Biofertilizers: a sustainable eco-friendly agricultural approach to crop improvement. *Plant Acclimation to Environmental Stress.* Tuteja, N., & Gill, S. S., (ed.), LLC 233 Spring Street, New York, 10013, USA: Springer Science plus Business Media, 403–432
38. Dogan, K., Kamail, C. I., Mustafa, G. M., & Ali, C., (2011). Effect of different soil tillage methods on rhizobial nodulation, biyomas and nitrogen content of second crop soybean. *Afr. J. Microbiol. Res., 5*, 3186–3194.
39. Aziz, G., Bajsa, N., Haghjou, T., Taule, C., Valverde, A., Mariano, J., & Arias, A., (2012). Abundance, diversity and prospecting of culturable phosphate solubilizing bacteria on soils under crop–pasture rotations in a no-tillage regime in Uruguay. *Appl. Soil Ecol., 61*, 320–326.
40. Choudhury, M. A., & Kennedy, I. R., (2004). Prospects and potentials for system of biological nitrogen fixation in sustainable rice production. *Biol. Fertil. Soils., 39*, 219–227. 10.1007/s00374–003–0706–2.
41. Revillas, J. J., Rodelas, B., Pozo, C., Martinez-Toledo, M. V., & Gonzalez, L. J., (2000). Production of B-Group vitamins by two Azotobacter strainswith phenolic compounds as sole carbon source under diazotrophicand adiazotrophic conditions. *J. Appl. Microbiol., 89*, 486–493.
42. Abd El-Fattah, D. A., Ewedab, W. E., Zayed, M. S., & Hassaneina, M. K., (2013). Effect of carrier materials, sterilization method, and storage temperature on survival and biological activities of *Azotobacter chroococcum* inoculants. *Ann Agric Sci., 58*, 111–118.
43. Sharma, P., Sardana, V., & Kandola, S. S., (2011). Response of groundnut (*Arachis hypogaea* L.) to rhizobium inoculation. *Libyan Agric. Res. Centre J. Int., 2*, 101–104.
44. Pindi, P. K., Satyanarayana, S. D. V., (2012). Liquid microbial consortium- a potential tool for sustainable soil health. *J. Biofertil. Biopest., 3*, 4.
45. Ambrosini, A., Beneduzi, A., Stefanski, T., Pinheiro, F., Vargas, L., & Passaglia, L., (2012). Screening of plant growth promoting Rhizobacteria isolated from sunflower *Helianthus annuus* L. *Plant & Soil, 356*, 245–264. 10.1007/s11104-011-1079-1.
46. Mohammadi, K., & Yousef S. Y., (2012). Bacterial biofertilizers for sustainable crop production: A review. *J. Agric. Biol. Sci., 7*, 307–316.
47. Hussain, N., Mujeeb, F., Tahir, M., Khan, G. D., Hassan, N. M., & Bari, A., (2002). Effectiveness of rhizobium under salinity stress. *Asian J. Plant. Sci., 1*, 12–14.
48. Ruiz-Sanchez, M., Aroca, R., Munoz, Y., Polon, R., & Ruiz-Lozano, J. M., (2010). The arbuscular mycorrhizal symbiosis enhances the photosynthetic efficiency and the antioxidative response of rice plants subjected to drought stress. *J. Plant Physiol., 167*, 862–869.

49. Tang, J., Wang, R., Niu, X., Wang, M., & Zhou, Q., (2010). Characterization on the rhizoremediation of petroleum contaminated soil as affected by different influencing factors. *Biogeosciences Discuss, 7*, 4665–4688. 10. 5194/bgd-7–4665–2010.

50. Ryu, C. M., Farag, M. A., Hu, C. H., Reddy, M. S., Kloepper, J. W., & Pare, P. W., (2004). Bacterial volatiles induce systemic resistance in Arabidopsis. *Plant Physiol., 134*, 1017–1026.

51. Riedlinger, J., Schrey, S. D., Tarkka, M. T., Hampp, R., Kapur, M., & Fiedler, H. P., (2006). Auxofuran, a novel substance stimulating growth of fly agaric, produced by the mycorrhiza helper bacterium Streptomyces AcH 505. *Appl. Environ. Microbiol., 72*, 3550–3557.

52. Singh, J. S., & Strong, P. J., (2016). Biologically derived fertilizer: A multifaceted bio-tool in methane mitigation. *Ecotoxicol. Environ. Saf., 124*, 267–276. doi:10.1016/j. ecoenv.2015.10.018. Epub 2015 Nov 11.

53. Hirsch, A. M., (2004). Plant–microbe symbioses: a continuum from commensalism to parasitism. *Symbiosis, 37*, 345–363.

54. Kerrie, F., David, B., & Naomi, C. S., (2014). Understanding and engineering beneficial plant–microbe interactions: plant growth promotion in energy crops. *Plant Biotechnology Journal, 12*(9), pp. 1193–1206.

55. Bapaume, L., & Reinhardt, D., (2012). How membranes shape plant symbioses: signaling and transport in nodulation and *arbuscular mycorrhiza. Front Plant Sci., 3*, pp. 223.

56. Bacilio-Jiménez, M., Aguilar-Flores, S., Ventura-Zapata, E., Pérez-Campos, E., Bouquelet, S., & Zenteno, E., (2003). Chemical characterization of root exudates from rice (*Oryza sativa*) and their effects on the chemotactic response of *endophytic bacteria. Plant Soil, 249*, 271–277.

57. Bottini, R., Cassán, F., & Piccoli, P., (2004). Gibberellin production by bacteria and its involvement in plant growth promotion and yield increase. *Appl. Microbiol. Biotechnol., 65*, 497–503.

58. Straub, D., Yang, H., Liu, Y., Tsap, T., & Ludewig, U., (2013). Root ethylene signalling is involved in *Miscanthus sinensis* growth promotion by the bacterial *endophyte Herbaspirillum frisingense* GSF30(T). *J. Exp. Bot. Nov., 64*(14), 4603–4615.

59. Sauter, M., (2015). Phytosulfokine peptide signalling. *J. Exp. Bot., 2, 66*(17), 5161–5169. Doi: 10.1093/jxb/erv071.

60. Onofre-Lemus, J., Hernández-Lucas, I., Girard, L., & Caballero-Mellado, (2009). ACC (1-aminocyclopropane-1-carboxylate) deaminase activity, a widespread trait in Burkholderia species, and its growth-promoting effect on tomato plants. *J. Appl. Environ. Microbiol., 75*(20), 6581–6590.

61. Frugier, F., Kosuta, S., Murray, J. D., Crespi, M., & Szczyglowski, K., (2008). Cytokinin: secret agent of symbiosis. *Trends in Plant Science, 13*, 115–120.

62. Barker, S. J., & Tagu, D., (2000). The roles of auxins and cytokinins in mycorrhizal symbioses. *Journal of Plant Growth Regulation, 19*, 144–154.

63. Claudia, K., (2014). Analysis of plant microbe interactions in the era of next generation sequencing technologies. *Front. Plant Sci., 21*. |https://doi.org/10.3389/fpls.2014.00216.

64. Dawson, J. O., (2008). "Ecology of actinorhizal plants." *Nitrogen-Fixing Actinorhizal Symbioses, 6*, Springer. pp. 199–234. doi: 10.1007/978–1–4020–3547–0_8.

65. José, A. S. Z., Pierre, H. C., Jean-Charles, Q., Henri-Philippe, H., Adam, K., & Pascal, R., (1998). Production of Sinorhizobium meliloti nod gene activator and Repressor flavonoids from *Medicago sativa. Roots Molecular Plant–Microbe Interactions, 11*, 784–794.

66. Wim D'Haeze, & Marcelle, H., (2002). Nod factor structures, responses and perception during initiation of nodule development. *Glycobiology, 12*(6), 79–105.

67. Herman, P. S., (2000). ROOT nodulation and infection factors produced by *Rhizobial* bacteria. *Annu. Rev. Microbiol., 54*, 257–88.

68. Murray, J. D., (2011). Invasion by invitation: rhizobial infection in legumes. *Mol Plant Microbe Interact, 24*(6), 631–639. doi: 10.1094/MPMI-08–10–0181.

69. Oláh, B., Brière, C., Bécard, G., Dénarié, J., Gough, C., (2005). Nod factors and a diffusible factor from *arbuscular mycorrhizal* fungi stimulate lateral root formation in *Medicago truncatula* via the DMI1/DMI2 signalling pathway. *Plant J., 44*(2), 195–207.

70. Daniel, J. G., (2004). Infection and invasion of roots by symbiotic, nitrogen-fixing rhizobia during nodulation of temperate legumes. *Microbiol. Mol. Biol. Rev., 68*(2), 280–300.

71. Allan Downie, J., (2014). Legume nodulation. *Science Direct. Current Biology, 24*(5).

72. Ott, T., Van Dongen, J. T., Günther, C., Krusell, L., Desbrosses, G., Vigeolas, H., Bock, V., Czechowski, T., Geigenberger, P., & Udvardi, M. K., (2005). Symbiotic leghemoglobin are crucial for nitrogen fixation in legume root nodules but not for general plant growth and development. *Curr. Biol., 15*(6), 531–535.

73. Sugano, S. S., Shimada, T., Imai, Y., Okawa, K., Tamai, A., Mori, M., & Hara-Nishimura, I., (2010). Stomagen positively regulates stomatal density in Arabidopsis. *Nature, 463*(7278), 241–244.

74. Kathleen, M., & Jos, V., (2000). The "oxygen paradox" of dinitrogen-fixing bacteria. *Biol. Fertil. Soils, 30*, 363–373.

75. Tarakhovskaya, E. R., Maslov, Yu, & Shishova, M. F., (2007). "Phytohormons in algae." *Russian Journal of Plant Physiology, 54*(2), 163–170.

76. Vob, U. Bishopp, A., Farcot, E., & Bennett, M. J., (2014). Modelling hormonal response and development. *Trends Plant Sci., 19*, pp. 311–319.

77. Ke, Q. Wang, Z., Ji, C. Y., Jeong, J. C., Lee, H. S., Li, H., Xu, B., Deng, X., & Kwak, S. S., (2015). Transgenic poplar expressing Arabidopsis YUCCA6 exhibits auxin-overproduction phenotypes and increased tolerance to abiotic stress. *Plant Physiol. Biochem., 94*, pp. 19–27.

78. Walz, A., Park, S., Slovin, J. P., Ludwig-Müller, J., Momonoki, Y. S., & Cohen, J. D., (2002). "A gene encoding a protein modified by the phytohormone indoleacetic acid." *Proc. Natl. Acad. Sci. USA, 99*(3), 1718–1723.

79. Flasiński, M., & Hąc-Wydro, K., (2014). Natural vs synthetic auxin: studies on the interactions between plant hormones and biological membrane lipids. *Environ Res., 133*, 123–134.

80. Fahad, S. Hussain, S., Bano, A., Saud, S., Hassan, S., Shan, D., Khan, F. A., Khan, F., Chen, Y. T., Wu, C., Tabassum, M. A., Chun, M. X., Afzal, M., Jan, A., Jan, M. T., & Huang, J. L., (2015). Potential role of phytohormones and plant growth-promoting rhizobacteria in abiotic stresses: consequences for changing environment. *Environ. Sci. Pollut. Res., 22*, pp. 4907–4921.

81. Iqbal, N. Umar, S., Khan, N. A., & Khan, M. I. R., (2014). A new perspective of phytohormones in salinity tolerance: regulation of proline metabolism. *Environ. Exp. Bot., 100*, pp. 34–42.

82. Kinoshita, N., Wang, H., Kasahara, H., et al., (2012). A-Ala Resistant3, an evolutionarily conserved target of miR167, mediates arabidopsis root architecture changes during high osmotic stress. *Plant Cell., 24*, 3590–3602.

83. O'Brien, J. A., & Benkova, E., (2013). Cytokinin cross-talking during biotic and abiotic stress responses. *Front. Plant Sci., 4*, pp. 451.

84. Kang, C., Cho, N., Kim, Y., & Kim, J., (2012). Cytokinin receptor-dependent and receptor-independent path ways in the dehydration response of *Arabidopsis thaliana*. *J. Plant Physiol., 169*, pp. 1382–1391.

85. Nishiyama, R. Watanabe, Y., Fujita, Y., Tien, L. D., Kojima, M., Werner, T., Vankova, R., Yamaguchi-Shinozaki, K., Shinozaki, K., Kakimoto, T., Sakakibara, H., Schmuelling, T., Lam-Son, P. T., (2011). Analysis of cytokinin mutants and regulation of cytokinin metabolic genes reveals important regulatory roles of cytokinins in drought, salt and abscisic acid responses, and abscisic acid biosynthesis. *Plant Cell., 23*, pp. 2169–2183.

86. Yamaguchi, S., (2008). Gibberellin metabolism and its regulation. *Annu. Rev. Plant Physiol., 59*, pp. 225–251.

87. Yan, J., Tsuichihara, N., Etoh, T., & Iwai, S., (2007). "Reactive oxygen species and nitric oxide are involved in ABA inhibition of stomatal opening." *Plant Cell Environ. 30*(10), 1320–5.

88. Sreenivasulu, N. Harshavardhan, V. T., Govind, G., Seiler, C., & Kohli, A., (2012). Contrapuntal role of ABA: does it mediate stress tolerance or plant growth retardation under long-term drought stress? *Gene, 506*, pp. 265–273.

89. Zhang, J. Jia, W., Yang, J., & Ismail, A. M., (2006). Role of ABA in integrating plant responses to drought and salt stresses. *Field Crops Res., 97*, pp. 111–119.

90. Goyal, K., Walton, L. J., & Tunnacliffe, A., (2005). "LEA proteins prevent protein aggregation due to water stress." *Biochemical Journal, 388*(1), 151–157.

91. Borovskii, G., Stupnikova, I., Antipina, A., Vladimirova, S., & Voinikov, V., (2002). "Accumulation of dehydrin-like proteins in the mitochondria of cereals in response to cold, freezing, drought and ABA treatment." *BMC Plant Biology, 2*(5), doi:10.1186/1471–2229–2–5.

92. Shi, Y., Tian, S., Hou, L., Huang, X., Zhang, X., Guo, H., & Yang, S., (2012). Ethylene signaling negatively regulates freezing tolerance by repressing expression of *CBF* and Type-A *ARR* genes in Arabidopsis. *Plant Cell., 24*, pp. 2578–2595.

93. Larkindale, J., Hall, D. J., Knight, M. R., & Vierling, E., (2005). Heat stress phenotypes of Arabidopsis mutants implicate multiple signaling pathways in the acquisition of thermo-tolerance. *Plant Physiol., 138*, pp. 882–897.

94. Clouse, S. D., & Sasse, J. M., (1998). "Brassinosteroids: Essential regulators of plant growth and development." *Annu. Rev. Plant Physiol. Plant Mol. Biol., 49*, 427–451. doi:10.1146/annurev.arplant.49.1.427.

95. Rao, S. S. R., Vardhini, B. V., Sujatha, E., & Anuradha, S., (2002). Brassinosteroids – new class of phytohormones. *Curr. Sci., 82*, 1239–1245.

96. Caño-Delgado, A., Yin, Y., Yu, C., Vafeados, D., Mora-Garcia, S., Cheng, J. C., Nam, K. H., Li, J., & Chory, J., (2004). "BRL1 and BRL3 are novel brassinosteroid recep-

tors that function in vascular differentiation in Arabidopsis." *Development (Cambridge, England), 131*(21), 5341–5351. doi:10.1242/dev.01403. PMID 15486337.

97. Vardhini, B. V., & Anjum, N. A., (2015). Brassinosteroids make plant life easier under abiotic stresses mainly by modulating major components of antioxidant defense system. *Front. Environ. Sci., 2*, pp. 1–16.

98. Shapiro, A. D., (2005). Nitric oxide signaling in plants. *Vitam Horm., 72*, 339–398.

99. Abdul, R. W., Michael, G. P., Mohd, Y. W., & Savarimuthu, I., (2011). *Plant Signal Behav., 6*(11), 1787–1792.

100. Reinbothe, C., Springer, A., Samol, I., & Reinbothe, S., (2009). "Plant oxylipins: role of jasmonic acid during programmed cell death, defense and leaf senescence." *The FEBS Journal, 276*(17), 4666–4681. doi:10.1111/j.1742–4658.2009.07193.x.

101. Farmer, E. E., & Ryan, C. A., (1990). "Interplant communication: airborne methyl jasmonate induces synthesis of proteinase inhibitors in plant leaves." *Proc. Natl. Acad. Sci. USA, 87*(19), 7713–7716. doi:10.1073/pnas.87.19.7713.

102. Shabir, H., Wani, V. K., Varsha, S., Saroj, K., S., (2016). Phytohormones and their metabolic engineering for abiotic stress tolerance in crop plants. *The Crop Journal, 4*(3), pp. 162–176.

103. Sarvajeet, S. G., & Narendra, T., (2010). Polyamines and abiotic stress tolerance in plants. *Plant Signal Behav., 5*(1), 26–33.

104. Huwei, S., Jinyuan, T., Pengyuan, G., Guohua, X., & Yali, Z., (2016). The role of strigolactones in root development. *Plant Signal Behav., 11*(1), e1110662.

105. Foo, E., & Davies, N. W., (2011). Strigolactones promote nodulation in pea. *Planta., 234*, pp. 1073–1081.

106. Naeem, M., Masroor, M., Khan, A., & Moinuddin, C., (2012). Triacontanol: a potent plant growth regulator in agriculture. *Journal of Plant Interactions, 7*(2), 129–142.

107. Ryan, C. A., (2000). "The systemin signaling pathway: differential activation of plant defensive genes." *Biochimica. et Biophysica. Acta., 1477*(1–2), 112–121. doi: 10.1016/S0167–4838(99)00269–1. PMID 10708853.

108. Orsini, F., Cascone, P., De Pascale, S., Barbieri, G., Corrado, G., Rao, R., Maggio, A., (2010). "System independent salinity tolerance in tomato: evidence of specific convergence of abiotic and biotic stress responses." *Physiologia. Plantarum. 138*(1), 10–21. doi:10.1111/j.1399–3054.2009.01292.x. PMID 19843237.

CHAPTER 11

TRANSGENIC PLANTS

ANJALI PRIYADARSHINI and PRERNA PANDEY

CONTENTS

11.1   INTRODUCTION

Transgenic plants have emerged as a promising candidate for the production of recombinant proteins with great potential in industry and clinical prospects. This novel system offers several advantages over conventional of using conventional bioproduction systems such as bacteria, yeast, and cultured insect and animal cells. The advantages can be listed as safe system, scale of production increases multi-fold, cost-effectiveness, and the ease of distribution and storage. Currently, plant systems, also referred to as bio-factories, are being utilized for the expression of various proteins, which includes potential vaccines and pharmaceuticals, through employing several adaptations of recombinant processes and utilizing the most suitable tools and strategies. This is known as plant molecular farming. The level of protein expression is a critical factor in plant molecular farming, and this level fluctuates according to the plant species and the organs involved. The production of recombinant native and engineered proteins is a complicated procedure that requires an inter and multidisciplinary effort involving a wide variety of scientific and technological disciplines, ranging from basic biotechnology, biochemistry, and cell biology to advanced production systems. A lot of effort needs to be put in to achieve the desired goals.

11.2 TOOLS FOR TRANSGENIC PLANTS

A transgenic plant contains a gene or genes that has been artificially inserted. Origin of the inserted gene sequence called the *transgene*, may be an unrelated plant, or a different species. The motive of inserting a combination of genes in a plant is to make it useful and productive in the following ways:

i.   higher yield;
ii.   improved quality;
iii. pest or disease resistance;
iv.   tolerance to heat, cold, and drought;
v.   expression of foreign proteins with industrial or pharmaceutical value;
vi.   economical alternative to fermentation-based production systems, for example, plant-made vaccines or antibodies (*plantibodies*) are especially remarkable, as plants are free of human diseases, thus reducing screening costs for viruses and bacterial toxins [1].

The first transgenic plants were reported in 1983. This has been followed by expression of many recombinant proteins in several important agronomic species of plants including tobacco, corn, tomato, potato, banana, alfalfa, and canola (Table 11.1). Initially, the choice of plant system was based on convenience and the need to evaluate genetic constructs quickly. Tobacco plants proved to fulfill these criteria and thus became the choice for such experiments during infancy. Foreign gene expression in plants can be accomplished either by stable integration of foreign DNA, which results in transformation of the nuclear genome, or by transient expression using modified plant viruses. Stable genomic integration is accomplished by introducing foreign DNA in the plant either by Agrobacterium T-DNA vectors or by direct means. The integration method has the advantage of permitting large scale cloning, maintenance of selected high-expressing genes and the ability to sexually cross transgenes to obtain multiple proteins expressed in the same plant [2]. On the other hand, transient expression using viral vectors is harder to initiate, because the viral vector must be inoculated into individual host plants, but gives a greater yield of protein.

11.2.1   VECTOR MEDIATED GENE TRANSFER

11.2.1.1   Agrobacterium tumefaciens

Agrobacterium tumefaciens contains a plasmid, which is a small circular piece of DNA having its own origin of replication and is replicated independent of nuclear material. This property makes it a suitable and attractive candidate to be used in genetically modifying plants. This plasmid is known as tumor-inducing (Ti) plasmid. When a wounded plant is exposed to *A. tumefaciens*, it integrates a stretch of its DNA, called transferred DNA (T-DNA), to the plant's genome [4]. The bacterium transfers its own T-DNA, but if the T-DNA is removed and replaced with another gene, *A. tumefaciens* can be used to introduce that gene into the plant genome, thereby providing a vector for scientists to engineer beneficial genes into plants. All the native T-DNA genes are not necessary for this process; thus, the unnecessary parts could be replaced by the gene of interest. The only requirement for inserted genes to get transferred to plants is the two repeated border sequences of 25 base pairs flanking the genes in the vector [5].

TABLE 11.1   Few Examples of Genetically Engineered Plants

Model Plants	Plants	Model Plants	Plants
Root plants	Carrot *(Daucus carota)*	Ornamental plants	Camation *(Dianthus caryophyllus)*
	Cassava *(Manihot esculenta)*		Orchids *(Cymbidium spp., Oncidium, Phalaenopsis)*
	Potato *(Solanum tuberosum)*		Rose *(Rosa hybrida)*
	Sweet potato *(Ipomoea batatas)*		Petunia *(Petunia hybrida)*
Turf grasses	Perennial ryegrass *(Lolium perenne)*	Industrial plants	Sunflower *(Helianthus annuus)*
	Bermuda grass *(Cynodon spp.)*		Indian Mustard *(Brassica juncea)*
	Switchgrass *(Panicum virgatum)*		Canola *(Brasicca napus)*
Tropic plants	Banana *(Musa spp.)*		Cotton *(Gossypium hirsutum)*
	Pineapple *(Ananas comosus)*	Legume plants	Alfalfa *(Medicago sativa)*
	Sugarcane *(Saccharum spp.)*		Beans *(Phaseolus spp.)*
	Citrus spp., coffee *(Coffea spp.)*		Soybean *(Glycine max)*
	Papaya *(Carica papaya)*		Pigeon pea *(Cajanus cajan)*
Woody species	Eucalyptus, Pine *(Pinus radiata)*		Peanut *(Arachis hypogaea)*
	Cork oak *(Quercus suber)*		Peas *(Pisum sativum)*
	Poplar *(Populus spp.)*		
	Rubber tree *(Hevea brasiliensis)*		Chickpea *(Cicer arietinum)*
Medicinal plants	Ginseng *(Panax ginseng)*	Cereal crop	Rice *(Oryza sativa)*
	Hemp *(Cannabis sativa)*		Maize *(Zea mays)*
Nuts and fruits	Blueberry *(Vaccinium corymbosum)*		Rye *(Secale cereale)*

TABLE 11.1 (Continued)

Model Plants	Plants	Model Plants	Plants
	Walnut *(Juglans spp.)*		Sorghum *(Sorghum bicolor)*
	Strawberry *(Fragaria X ananassa)*		Wheat *(Triticum aestivum)*
	Apple *(Malus X domestica)*		Barley *(Hordeum vulgare)*
Vegetable plants	Tomato *(Lycopersicum esculentum)*	Model plants	Arabidopsis *(Arabidopsis thaliana)*
	Lettuce *(Lactuca sativa)*		Tobacco *(Nicotiana tobaccum. N. benthamiana)*

11.2.1.2 Binary Vectors

Other vectors can also be used to perform the transformation. Binary vectors, like the pBIN20 vector (Figure 11.1), are plasmids that contain the 25 base pair border sequences, allowing the new genes within them to be integrated into plant genomes, as well as marker genes that are used later in the process to select for successful gene transfer. The vectors also

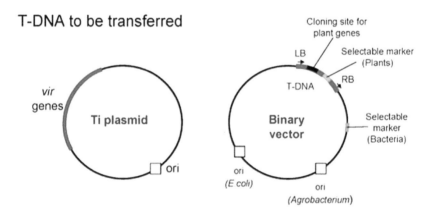

FIGURE 11.1 Binary vector system to develop transgenic plant, vir, and T-DNA can be on separate plasmids and only left and right borders (LB and RB) are required for T-DNA to be transferred.

contain origins of replication for *A. tumefaciens* and *Escherichia coli* [5]. Therefore, the plasmids can replicate themselves in either *E. coli* or *A. tumefaciens*, making possible for scientists to work with *E. coli* and then transfer the vectors to *A. tumefaciens* through bacterial conjugation, the process by which two bacteria exchange genetic information in the form of plasmids, when the genes are ready to be inserted into the plant genome.

11.2.1.3   Bacterial Artificial Chromosome

The limitation of binary vector of inability to introduce larger genes into plants has been mitigated by using bacterial artificial chromosomes (BACs). BACs are synthesized gene vectors based on a plasmid from *E. coli* [6]. BACs can have inserts ranging from 50 to 350 kb, allowing for the transfer of large genes or many small genes at a time. The BACs used to transform plant cells are binary vectors, termed binary-bacterial artificial chromosomes (BIBACs) [7]. Like binary vectors, BIBACs contain gene markers as well as the two flanking boundary regions. Two commonly used gene markers in BIBACs and other binary vectors are neomycin phosphotransferase II (NPTII), which confers resistance to the antibiotic kanamycin, and hygromycin phosphotransferase, which gives resistance to the antibiotic hygromycin.

Transgenes can be inserted between the boundary regions of the BIBAC or other binary vectors. Firstly, the restriction enzymes cleave DNA at specific sites within a sequence of base pairs. An enzyme that cuts with sticky ends can be used to insert a foreign gene into the vector of choice. This is accomplished by cutting both with the enzyme and then "gluing" the gene fragment to the vector using a DNA ligase that helps hydrogen bond the complementary bases of the sticky ends. Once the appropriate vector is made and transferred to the *A. tumefaciens* bacterium, it must be integrated into the plant genome. The leaf disks start growing shoots, which are the precursors of the plant's stem and leaves. Only those plant cells that have integrated T-DNA from the vector will have antibiotic resistance. The medium therefore selects for the transformed plant cells by killing those that do not contain the vector. Once the transformed cells are selected, the growing shoots are transferred to a root-inducing medium, where they grow roots, and then to soil to grow into transgenic plants.

11.2.2   DIRECT GENE TRANSFER METHODS

In the direct gene transfer methods, the foreign gene of interest is delivered into the host plant cell without the help of a vector. The methods used for direct gene transfer in plants are:

11.2.2.1   Chemical Mediated Gene Transfer

Chemicals like polyethylene glycol (PEG) and dextran sulfate induce DNA uptake into plant protoplasts. Calcium phosphate is also used to transfer DNA into cultured cells.

11.2.2.2   Microinjection

DNA is directly injected into plant protoplasts or cells (specifically into the nucleus or cytoplasm) using fine tipped) glass needle or micropipette. This method of gene transfer is used to introduce DNA into large cells such as oocytes, eggs, and the cells of early embryo.

11.2.2.3   Electroporation

This involves a pulse of high voltage applied to protoplasts/cells/ tissues to make transient (temporary) pores in the plasma membrane, which facilitates the uptake of foreign DNA. The cells are placed in a solution containing DNA and subjected to electrical shocks to cause holes in the membranes. The foreign DNA fragments enter through the holes into the cytoplasm and then to nucleus.

11.2.2.4   Particle Gun/Particle Bombardment

In this method, the foreign DNA containing the genes to be transferred is coated onto the surface of minute gold or tungsten particles (1-3 micrometers) and bombarded onto the target tissue or cells using a particle gun (also called as gene gun/shot gun/microprojectile gun). Two types of plant tissue are commonly used for particle bombardment: primary explants and the proliferating embryonic tissues.

11.2.2.5   Transformation

This method is used for introducing foreign DNA into bacterial cells, e.g., *E. coli*. The transformation frequency (the fraction of cell population that can be transferred) is very good in this method. For example, the uptake of plasmid DNA by *E. coli* is carried out in ice cold $CaCl_2$ (0–5°C) followed by heat shock treatment at 37–45°C for about 90 s. The transformation efficiency refers to the number of transformants per microgram of added DNA. $CaCl_2$ breaks the cell wall at certain regions and binds the DNA to the cell surface.

11.2.2.6   Conjugation

It is a natural microbial recombination process and is used as a method for gene transfer. In conjugation, two live bacteria come together and the single-stranded DNA is transferred via cytoplasmic bridges from the donor bacteria to the recipient bacteria.

11.2.2.7   Liposome-Mediated Gene Transfer or Lipofection

Liposomes are circular lipid molecules with an aqueous interior that can carry nucleic acids. Liposomes encapsulate the DNA fragments and then adhere to the cell membranes and fuse with them to transfer DNA fragments. Thus, the DNA enters the cell and then to the nucleus. Lipofection is a very efficient technique used to transfer genes in bacterial, animal, and plant cells.

11.3   PLANT-CELL SUSPENSION CULTURES

One of the best plant-based alternatives to mammalian cells for the production of biopharmaceuticals is a plant-cell suspension, which is a robust system involving a simple purification procedure and easy downstream processing [8]. Although this system requires a high level of sterility to control contamination, pharmaceuticals with a high level of purity can be produced using it [9]. Additionally, in contrast to plant-cell culture systems, plant-cell suspension system eliminates the regeneration process, and accordingly, is a rapid procedure. Despite having numerous advantages over other systems, plant-cell suspension system has been established only for a small number of plants, such as tobacco, carrot, *Arabidopsis*, and rice. Moreover, due to the certain limiting factors, such as increasing proteolytic activity, which leads to a low concentration of the recombinant protein during the late stationary phase, this system is not the best method of protein expression [10].

11.4   METHODS OF TRANSGENIC PLANT PRODUCTION OTHER THAN NUCLEAR TRANSFORMATION

11.4.1   TRANSPLASTOMIC

Plants are novel alternatives to nuclear transgenic plants, which are created by introducing the recombinant DNA into the genome of chloroplasts rather than the nuclear genome using gene-gun bombardment. Some of the most important accessible products that have been produced in tobacco chloroplasts are a tetanus-toxin fragment, serum albumin, and human growth hormone. However, plastids do not have the capacity to perform glycosylation; thus, chloroplast cannot be used to produce human glycoproteins [11].

11.4.2   PLASTID TRANSFORMATION

Plastid transformation is an efficient alternative to nuclear transformation because it has several advantages that the latter method lacks. For example, despite the enormous importance of the delivery of a normal bio-containment of transgene flow by out-crossing, the transgene cannot be transferred due to the lack of chloroplasts in pollen, thereby allaying public concerns regarding genetically modified plants. Transgenic plants subjected to homoplastomic chloroplast transformation are selected after several generations of plants have been regenerated from the gene-gun bombarded leaf explants, meaning that the plant chloroplast genome has had opportunity to incorporate the transgene. The selection of the above-mentioned bombarded leaf explants is conducted on a medium containing a either spectinomycin or streptomycin.

Researchers have achieved a noticeable yield of therapeutic human and bacterial proteins, ranging from 3–6% of the total soluble proteins by using the tobacco chloroplast-transformation technique [12]. High expression of antibiotics which are proteinaceous in nature comprising of 70% soluble protein has been obtained using this system. This transgenic antibiotic protein from transgenic plant has been a very significant development. Although plastid transformation has an enormous potential, its application remains restricted. In that regard, even though plastid transformation has been attained in plant species such as lettuce, eggplant, soybeans and tomatoes, the plant most commonly modified using a chloroplast transformation system is tobacco, which is highly regulated and is inedible due to its high level of toxic alkaloids. Finally, whether the protein stability will change over time, even with refrigeration, is a matter of concern.

11.5 PLANT MOLECULAR FARMING VERSUS CONVENTIONAL BIOPRODUCTION SYSTEMS

Compared with other transgenic products obtained from transgenic bacteria, fungi, and animals, which are the most common models for recombinant-protein production till date, transgenic plant product have evolved as a very attractive and viable tool for large scale production.

11.5.1 ECONOMIC VIABILITY

Plant-based therapeutics, which are fast replacing chemical therapeutics, are produced at a low budget, thus providing an economic justification for their use [13, 14] and all the effort and resource put into its development.

11.5.2 EASE OF OPERATION

Plant molecular farming can harness the availability of personnel with experience and expertise in planting, harvesting, instead of highly trained personnel and well-equipped laboratories as an essential requirement for the conventional procedures. Processing of plant material for recombinant protein production too adds to the advantage as it is much simpler. Another beneficial characteristic of these systems is that recombinant plants can be stored at room temperature. In contrast, the storage temperature for plant viruses, bacteria and yeasts is $-20°C$. The storage condition for cultured mammalian cells is even more stringent because they must be maintained in liquid nitrogen.

11.5.3   MINIMIZE THE HEALTH-ASSOCIATED RISK

In addition, the dissimilarity of the pathogenic factors of plants and humans decreases the risk of microbial interactions that could negatively affect the quality of the final products.

11.5.4   ENHANCE STABILITY OF THE PRODUCT

Technically, the stability of recombinant proteins within plants stressed by environmental factors is greater than that of recombinant proteins produced in other hosts. Furthermore, higher plants typically produce recombinant proteins with the correct folding, activity, and glycosylation [15].

11.5.5   GENERATION OF VIRUS FREE PLANT

If viral transgenes are introduced and silenced, the post-transcriptional process also prevents homologous RNA viruses from accumulating; this is a means of generating virus-resistant plants.

11.6   DISADVANTAGES OF PLANT MOLECULAR FARMING

i.   There are limits to the products than can be obtained using plant molecular farming, such as the unknown mechanisms that cause certain post-translational disorders in plant cells.

ii. The challenging issue of how to fine-tune the systems that are essential for the preservation of the structural integrity of the nascent recombinant proteins and their activities in their new cellular environments are still debated within the field of plant molecular farming. The advantages and disadvantages of plant molecular farming compared with using other molecular farming systems are presented in.

iii. In genetically modified plants, the introduced transgenes are sometimes not expressed. They can be silenced too, this occurs due to phenomenon of co-suppression when transgene are homologous to plant endogenous genes. Silencing occurs transcriptionally and post-transcriptionally, but silencing of endogenous genes seems predominantly post-transcriptional.

11.7   IMPORTANT TRANSGENIC PLANTS

11.7.1   TOBACCO

One of the most suitable platforms typically used as a green producer is tobacco (*Nicotiana tabacum*). It has many unique advantages over other plant species for the production of pharmaceutically important proteins which are:

i.   This herb is capable of producing a biomass of up to 100 ton/ha.
ii. Moreover, a well-established system for transforming tobacco that results in a high level of soluble protein exists.
iii. The potential of utilizing various strategies for the expression of proteins in a stable or transient manner using this species, as well the possibility of using chloroplast genome-based methods mediated via *Agrobacterium* or viral induction are its other advantages [16].
iv. Tobacco is not a feed or food crop, which decreases the probability of its contamination within the feed or food chains.

Disadvantages

i.   High concentrations of alkaloids and nicotine in some tobacco varieties are disadvantageous to utilize this plant for molecular farming.

ii. The instability of the products of tobacco is one of the disadvantages of this plant in molecular farming.

Tobacco has been shown to produce a wide variety of therapeutic immune-modulatory molecules, such as cytokines, vaccines, and antibodies [17]. To date, various types of these vaccines have been generated in tobacco, and some of them have been tested in mammals.

Besides the therapeutic molecules, Tobacco can be modified to produce cytokines, some of which are glycoproteins. cytokines are immuno-regulators produced by different types of cells in an animal cell. Cytokines are involved in regulating the immune response majorly by inhibiting or stimulating the activation, proliferation, and differentiation of a variety of immune cells. Though cytokines have been produced in *E. coli*, tobacco has emerged as a very good alternative to overcome the limitations of some conventional bioproduction systems, which are low levels of expression and the lack of glycosylation of *E. coli*-derived recombinant cytokines.

11.7.2 ARABIDOPSIS

Arabidopsis has been among the main plants [18] used for genetic transformation due to its various characteristics such as:

i. the short generation period;
ii. small genome size;
iii. presence of a self-pollination mating system;
iv. ease of *in vitro* culturing, easy regeneration, and *in vivo* transformation;
v. its lack of food and feed applications.

11.7.3 CEREALS AND LEGUMES

Seeds have been utilized greatly for molecular farming because of the fact that the double up as protein synthesis and protein storage organelles. A seed-based system has an added advantage for molecular farming because they allow the long-term storage of proteins which is facilitated by adequate biochemical environment, a low water content, and low protease activities and last but not the least they possess biosafety and are easy to transport. The disadvantage of such process is that the seed-based transgenic plants need to reach the flowering stage before the recombinant proteins can be extracted. This increases the possibility of environmental contamination by the pollen of the transgenic plants. List of some of recombinant proteins, extracted from seed bioreactors include vaccine antigens, cell-culture proteins, industrial enzymes, therapeutic antibodies and cytokines.

Among the seed-based bioreactors, maize is the major viable plant that produces recombinant proteins in large amounts. Some of the remarkable features regarding this model plant compared with those of other plant systems used in molecular farming are the existence of well-established techniques for its tissue culture and transformation, its production of high levels of biomass and the ease of scaling up its use. Moreover, this plant has the ability to produce recombinant antibodies as well as protease inhibitors and enzymes for pharmaceutical/technical applications, such as aprotinin, laccase, and trypsin [19].

Successful expression of human lactoferrin was achieved using rice as the host plant [20]. The recombinant single-chain Fv antibody directed against carcinoembryonic antigen that was produced in wheat and rice could be preserved for up to 4 or 5 months at room temperature without any loss in activity or of the product [21] which is of great value in storage and transportation of the product.

Soybean and alfalfa plants have the ability to directly utilize atmospheric nitrogen through nitrogen fixation, thus making them ideal plants for the production of recombinant antibodies and other proteins. The drawback of this plant system is that it produces a relatively smaller amount of green biomass than that by tobacco and maize.

11.7.4 VEGETABLES AND FRUITS

Potatoes have been put to extensive use for the production of plant-derived vaccines, which have been administered to humans in many clinical trials. Potato (*Solanum tuberosum*), banana (*Musa acuminata colla*), tomato

(*Lycopersicon esculentum*), and carrot (*Daucus carota* spp. *sativus*) plants have been successfully utilized for the expression of vaccine subunits. Interestingly, among these plants, tomato plants have been utilized as model genetically transformed producers of the first plant derived rabies-vaccine component, HIV-gag, and HBsAg proteins [22]. Alzheimer's disease (AD), which is a neurodegenerative disease, progresses through accumulation of beta amyloid (Aβ) in the brain, leading to neuronal destruction and intensification of the disease process. A strategy to fight this could be administration of an antigen directed against this toxic protein, which might prove to be the most useful to treat AD or in the least to arrest its progression.

Scientists have been able to successfully express Aβ in tomatoes [23]. Another protein thymosin α1 has been expressed in tomato fruits through *Agrobacterium*-mediated transformation and been used to treat cancer and viral infections [24]. Ma, et al. [25] demonstrated that the potato tuber is a suitable host for the production of diagnostic antibody-fusion proteins, human milk proteins, and other antibodies. Banana plants too have been considered as green bioreactors for the production of recombinant antibodies and vaccines. The advantage of transgenic fruits of different varieties of banana commonly grown in countries in which vaccines are most desperately needed is that it can be consumed as raw materials by both children and adults or the vaccine subunits within them can be purified [25].

11.7.5   OILSEEDS

Plants bearing seeds that are rich in oil are known as oilseed plants, for example, safflower, coconut, peanut, sunflower, palm, sesame, olive, rapeseed, and rice (bran) to name a few. Major advantage associated with the use of oil seed is that they are useful sources of recombinant proteins due to their protein-production capability along with the ease of purifying the proteins produced. Additional benefits are the low cost and low acreage associated with their use, the high protein yield obtained, and being self-pollinating plants. Oleosins are small structural proteins that are attached to the surface of oil bodies and subcellular organelles that store oils. Oleosin-recombinant fusion proteins must be detached so that they can be extracted, which can be achieved using a simple procedure, namely, endoprotease digestion. Safflower-derived insulin and hirudin are new generation pharmaceutical proteins that have produced by oil-seed plants [26].

11.7.6 AQUATIC PLANTS

Aquatic plants have also emerged as promising green-cell platforms for the introduction of genes and the production of novel recombinant proteins. For this, the chloroplast and nuclear genomes of five different microalgal species, namely *Amphidinium carterae*, *Pheaeodactylum tricornutu*, *Cylindrotheca fusifornzis*, *Symbiodinium microadriaticum*, and *Chlamydomonas reinhardtii*, have been successfully transformed. Advantage of this system are:

    i.   lack of toxicity;
    ii.   low expression levels;
    iii. low cost;
    iv. short growth period;
    v.   high yield;
    vi. capability for fresh yield.

11.8   ADVANTAGES ASSOCIATED WITH TRANSGENIC PLANTS

11.8.1   GLYCOSYLATION AS A MODIFICATION MECHANISM IN TRANSGENIC PLANTS

Post-translational modifications such as glycosylation, phosphorylation, sulfation, and methylation are vital in transgenic higher eukaryotic organisms. Modifying recombinant proteins through post-translational processes plays an important role in their functions. One of the most important post-translational modifications is glycosylation. Certain aspects of this enzymatic process, in which glycans are attached to organic molecules such as proteins and lipids, are significantly different in plants and mammalian cells. The basic N-acetyl glucosamine (GlcNAc)-mannose precursor structures added to the glycosylation sites of proteins within the endoplasmic reticulum of plant and mammalian cells are identical, but evolution of the Golgi apparatus has caused considerable variation. Glycosylation affects the basic biological functions of proteins, such as their ligand receptor interactions, specific activity and immunogenicity. A chain of oligosaccharides can be formed via either O- or N-linked glycosylation. N-linked glycosylation occurs within the endoplasmic reticulum (ER), and the primary oligosaccharide chain is further processed during its exit from the ER and passage through the Golgi apparatus (GA).

Unfortunately, the products of plant glycosylation can occasionally lead to side effects, such as allergic reactions. Because persons who are prone to pollen allergies have IgE (Ig: insoluble glycoproteins) and IgG4 reactivity to glycoepitopes, it is rational to attribute the allergenicity of plant-based glycosylated antibodies to the existence of allergenic factors, such as glycoproteins containing $\alpha(1,3)$-fucose and $\beta(1,2)$-xylose.

11.8.2   LARGE-SCALE TRANSIENT GENE TRANSFECTION, CLIMATE RISK FREE PRODUCTION SYSTEMS, AND BIOSAFETY CONSIDERATIONS

Plant biotechnology typically relies on two strategies for delivery and expression of heterologous genes in plants, including: (a) stable genetic transformation, and (b) transient expression using viral vectors. In recent years, the technological progression in virus-based vectors has allowed plants to become a feasible platform for recombinant proteins (RPs) production, while RPs were only able to be produced from cultures of mammalian, insect, and bacteria cells, previously. The plant-based RPs are more preferable in terms of versatility, speed, cost, scalability, and safety.

In spite of being a faster method, the transient approach is hampered by low contagiosity of viral vectors carrying average- or large-sized genes. Fortunately, these drawbacks have been subject to troubleshooting by developing constructs for the efficient delivery of RNA viral vectors as DNA precursors. The mentioned efforts have tended to expanding systemic *Agrobacterium tumefaciens*-mediated transfection of viral replicons for efficient transient expression in plants. As such, *Agrobacterium*-mediated delivery of the target constructs using results in gene amplification in all developed leaves of a plant simultaneously. This process is also referred to as "magnifection" that can be performed on a large scale and with different plant species. The mentioned technique incorporates advantages of three biological systems consisting of:

  a) the transfection efficiency of *A. tumefaciens*,
  b) the high expression yield obtained with viral vectors, and
  c) the post-translational capabilities of a plant.

Transient expression systems have been established to eliminate the long-time frame of generating transgenic plants, so that the transgene is not integrated into the plant genome but rather quickly directs the production of the RP while residing transiently within the plant cell. In addition to the significant acceleration of production timeline, this approach improves the recombinant proteins accumulation level by excluding the "position effect" of variable expression instigated by the random integration of transgene within the genome. In other words, the climate risk free molecular farming systems have become more achievable by conducting the transient gene transfection. Beside all these advances achieved by the transient expression technology, some complementary strategies have been taken into consideration to limit the potential environmental and human health impacts linked to PMF. Specifically, cell cultures of transgenic plants, physical containment, dedicated land, plastid transformation, biological confinement, male sterility, gene use restriction technologies (GURTs), expression from or in roots, expression in edible parts and seeds, post-harvest inducible expression, and temporal confinement have been suggested as additional solutions to minimize the risks of PM.

11.8.3   INDUSTRIALIZATION, CURRENT STATUS, AND PERSPECTIVES

As plant molecular farming has come of age, there have been technological progresses on many aspects, including transformation methods, regulating gene expression, protein targeting and accumulation, as well as the use of different crops as production platforms. Recently, plant molecular farming has been proposed as an example of a green development scheme in convergence with sustainable agricultural industries. Despite this progress, yield improvement remains as one of the most challenging issues, because the product yield has a significant impact on economic feasibility of any related project. The advantages of transgenic plants over other expression systems make them become industrialized as economic alternatives to the conventional pharmaceutics. Several plant-made pharmaceuticals, including the enzyme glucocerebrosidase (GCase), insulin and Interferon alfa 2b [IFN-alpha (2b)], have approached commercialization with low costs and large-scale production. Interestingly, these achievements have been attached to substantial patenting activities as well.

11.9 CONCLUSIONS

Plant molecular farming has been shown to be a promising biotechnological approach; however, because this approach is novel, its efficacy may be disputed. Methods that facilitate plant cultivation under extremely controlled conditions should be developed for the subsequent stages of this process, as we move away from aseptic plant-cell cultures to non-aseptic conditions in which plants are grown traditionally or are grown hydroponically using compost. Plant molecular farming has significant potential for the development of medicinal products. With regard to the history of plant molecular farming, the current major focus is to accelerate the improvement of plant biotechnological procedures for the generation of new products, as well as conventional products. The most important challenges in this field are identifying new plant resources and optimizing protocols for producing high levels of recombinant proteins.

11.10 REVIEW QUESTIONS

1. What is plant molecular farming?
2. Describe *Agrobacterium*-mediated gene transfer in plant cell.
3. What is a double vector system?
4. What are the various nonvector-associated techniques to form transgenic plant?

5. What are the advantages of using plant-based bioreactor?
6. Why has tobacco been proved to be a very important experimental model?
7. What are the various recombinant proteins obtained from transgenic plants?

KEYWORDS

- nuclear transformation
- plant molecular farming
- suspension culture
- transgenic plant

REFERENCES

1. Herbers, K., Grenier, G., et al., (1999). Ectopic expression of a tobacco invertase inhibitor homolog prevents cold-induced sweetening of potato tubers, *Nature Biotechnology, 17*, 708–711.
2. Tacket, C. O., et al., (1998). Immunogenicity in humans of a recombinant bacterial antigen delivered in a transgenic potato, *Nat. Med.*, 607–609.
3. Attikum, H. V., Paul, B., et al., (2001). Non-homologous end-joining proteins are required for *Agrobacterium* T-DNA integration, *The EMBO Journal, 20*(22), 6550–6558.
4. St Schell J., (1987). Transgenic plants as tools to study the molecular organization of plant genes, *Science, 237*(4819), 1176–83.
5. Walden, R., Koncz, C., & Schell, J., (1990). The use of gene vector in plant molecular biology, *Methods in Molecular and Cellular Biology, 1*, 175–193.
6. Griffith, A. J. F., (2000). *An Introduction to Genetic Analysis*. 7th edition.
7. Hamilton, C. M., Frary, A., et al., (1999). Construction and characterization of BIBAC genomic DNA libraries of *Lycopersicon esculentum* and *Lycopersicon pennelli*. *The Plant Journal, 1*, 223–229.
8. Kim, T. G., Baek, M. Y., et al., (2008). Expression of human growth hormone in transgenic rice cell suspension culture. *Plant Cell. Rep., 27*, 885–891.
9. Franconi, R., Dmurtas, O. C., et al., (2010). Plant derived vaccines and other therapeutics produced in contained systems, *Expert Rev. Vaccine, 9*, 877–892.
10. Obembe, O. O., Popoola, J. O., et al., (2011). Advances in plant molecular farming, *Biotechnol. Adv., 29*, 210–222.

11. Ma, W., Zhang, W., et al., (2003). Multiple PCR typing of high molecular weight glutenin alleles in wheat, *Euphytica, 134*, 51–60.

12. Reddy, G. S. N., Prakash, J. S. S., et al., (2002). Arthrobacter roseus sp. nov., a psychrophilic bacterium isolated from an Antarctic cyanobacterial mat sample, *Int. J. Syst. Evol. Microbiol., 52*, 1017–1021.

13. Häkkinen, S. T., Raven, N., et al, (2014). Molecular farming in tobacco hairy roots by triggering the secretion of a pharmaceutical antibody, *Biotechnol. Bioeng., 111*, 336–346.

14. Manoj, M. K., Ahwari, H., et al., (2106). Tackling unwanted proteolysis in plant production hosts used for molecular farming, *Front. Plant Sci., 7*, 267.

15. Yano, M., Hirai, T., et al., (2010). Tomato is a suitable material for producing recombinant miraculin protein in genetically stable manner, *Plant Sci., 178*, 469–473.

16. Saksia, K. R., Pauli, K. T., et al., (2009). The production of biopharmaceuticals in plant systems, *Biotechnology Advances, 27*(6), 879–894.

17. Tremblay, M., Richard, L., et al., (2014). Learning reflexively from a health promotion professional development program in Canada. Health Promotion International, *Advance Access, 29*, 538–548.

18. Koornneeef, M., & Meinke, D., (2010). The development of arabidopsis as a model plant, *Plant J. 61*(6), 909–21.

19. Hood, E. E., Bailey, M. R., et al., (2003). Criteria for high level expression of a fungal laccase gene in transgenic maize, *Plant Biotechnol. J., 1*, 129–140.

20. Anzai, H., Takaiwa, F., et al., (2000). Production of human lactoferrin in transgenic plants, Elsevier, *Amsterdam*, 265–271.

21. Stoger, E., Vaquero, C., et al., (2000). Cereal ropsas viable production and storage systems for pharmaceutical scFv antibodies, *Plant Mol. Biol., 42*, 583–590.

22. Sala, F., et al., (2003). Vaccine antigen production in transgenic plants: gene constructs and perspective, *Vaccine*, 803–808.

23. Youm, J. W., et al., (2008). Transgenic tomatoes expressing human beta amyloid for use as a vaccine against Alzheimer's disease, *Biotecnol. Lett., 30*, 1839.

24. Cheng, C. P., & Nagy, P. D., (2003). Mechanism of RNA recombination in carmo and tomb viruses: evidence for template switching by the RNA-dependent RNA polymerase *in vivo, J. Virol., 77*, 12033–12047.

25. Ma, J. K. C., Drake, P. M. W., et al., (2003). The production of recombinant pharmaceutical proteins in plants, *Nat. Rev. Genet., 4*, 794–805.

26. Spok, A., & Karner, S., (2016). *Plant Molecular farming- Opportunities and Challenges*, European Commission, Institute for Prospective Technological Studies, Seville, http://ftp.jrc.es/EURdoc/JRC43873.Pdf.Zugegriffen:Februar.

CHAPTER 12

GLOBAL FOOD SAFETY

PRERNA PANDEY and ANJALI PRIYADARSHINI

CONTENTS

12.1 INTRODUCTION

Wendell Berry quoted, "eating is an agricultural act." According to Food and Agriculture Organization (FAO), the view that "eating is also an ethical act" resonates well (FAO, 2013) developed with the problem of malnutrition sweeping across the world.

Food ethics raises issues and asks questions in relation to food all along the value chains. It also puts things into relation with each other. Human civilization has always been associated with the field of food safety as one of the prime goals has been increasing the durability of food.

A huge global food chain has emerged in this field. For instance, a food item like fish or a plant part may be grown by a poor farmer to be consumed by a rich consumer in another corner of the world. In between these two who may not be aware of each other's existence are a slew of middlemen.

The poignant statement on food security, "Food security exists when all people, at all times, have physical and economic access to sufficient, safe and nutritious food that meets their dietary needs and food preferences for an active and healthy life," was made by FAO in 1996 [1].

Food safety has been identified at domestic, regional, and international levels as a public health priority, as unsafe food causes illness in millions of people every year and many deaths. According to the World Health Organization (WHO), serious outbreaks of foodborne disease have been documented on every continent in the past decade, and in many countries, the rates of related illnesses are increasing significantly. Key global food safety concerns include the spread of microbiological hazards (including bacteria such as *Salmonella* or *Escherichia coli*, etc.); chemical food contaminants; assessments of new food technologies (such as genetically modified food); and strong food safety systems in most countries to ensure a safe global food chain.

The WHO enlists several aspects of food safety like how contaminated food may cause long- term health issues; owing to the complexity of today's global food chain, there are several stages where contamination may be introduced. WHO recognizes that food safety is essential given today's global scenario and that consumers must be aware of food safety.

12.2 FOOD AND AGRICULTURE ORGANIZATION (FAO)

FAO works with governmental authorities, local industry, and other relevant stakeholders to ensure food safety to consumers. An estimated three million people around the world, in developed and developing countries, die every year from food and water-borne disease, with millions more becoming sick. Occurrence of such disease can easily escalate to a food safety emergency situation, which can adversely impact national economies and livelihoods through reduced availability of food for national consumption, closure of export markets, and/or the high cost of addressing the effects of the threat.

To contribute to the efforts to reduce this adverse impact of food safety emergencies on global food security and public health, and at the request of its members, The Food and Agriculture Organization of the United Nations (FAO) has established an Emergency Prevention System for Food Safety (EMPRES Food Safety). EMPRES Food Safety will complement and enhance FAO's ongoing work in food safety as well as in animal health and plant health emergencies.

12.3 THE GLOBAL FOOD SAFETY INITIATIVE (GFSI)

The Global Food Safety Initiative (GFSI) is an industry-driven initiative providing thought leadership and guidance on food safety management systems necessary for safety along the supply chain. This work is accomplished through collaboration between the world's leading food safety experts from retail, manufacturing, and food service companies as well as international organizations, governments, academia, and service providers to the global food industry. They meet together at technical working group and stakeholder meetings, conferences, and regional events to share knowledge and promote a harmonized approach to manage food safety across the industry. GFSI is facilitated by The Consumer

Goods Forum (CGF), a global, parity-based industry network, driven by its members.

The objectives of the GFSI may be summarized as reduce food safety risks, manage the cost of food production, and enhance the efficacy of the system (Source: http://www.mygfsi.com/about-us/about-gfsi/what-is-gfsi.html) [2].

12.4   THE BRITISH RETAIL CONSORTIUM (BRC)

BRC was the first to be recognized as meeting the GFSI benchmark; though it originated in the UK, there are many BRC accrediting centers across 90 countries.

The FSSC 22000 Food Safety System Certification provides a framework for effectively managing an organization's food safety responsibilities (Source: https://www.cert-id.com/Certification-Programs/BRC-Certification.aspx) [3]. FSSC 22000 is fully recognized by the Global Food Safety Initiative (GFSI) and is based on existing ISO Standards. The FSSC 22000 Scheme sets out the requirements for certification bodies (CBs) to develop, implement, and operate a certification scheme and to guarantee its impartiality and competence. The scheme sets out the requirements to assess the food safety system of food manufacturing organizations and to issue a certificate. This certificate indicates that the organizations food safety system is in conformance with the requirements that are given in the scheme and that the organization is able to maintain conformance with these requirements. The value added to an organization with a certified food safety system lies in the efforts made by the organization to maintain that system and its commitment to continuously improve its performance.

12.5   THE INTERNATIONAL ASSOCIATION FOR FOOD PROTECTION (IAFP)

IAFP represents a broad range of members with a singular focus — protecting the global food supply (*Source:* https://www.foodprotection.org/) [4].

12.6   THE INTERNATIONAL UNION OF FOOD SCIENCE AND TECHNOLOGY (IUFOST)

IUFoST is a global food security organization with 75 member countries (Source: http://www.iufost.org/about-iufost) [5]. It organizes food congresses and other activities stimulating knowledge in the fields of expansion, distribution, and conservation of food resources.

WHO Estimates of the Global Burden of Foodborne Diseases published in December 2015 [6] explains the diseases that people get from eating contaminated food are an important cause of illness, disability, and deaths around the world. Foodborne diseases, especially those caused by bacteria, viruses, parasites and fungi, are preventable, and education in safe food handling is a key measure for protection.

12.7   THE FIVE KEYS TO SAFER FOOD PROGRAM

WHO built the five keys to Safer Food Program to assist member states in promoting safe food handling activities and educate all food handlers, including consumers, with tools easy to adopt and adapt. The Five Keys to Safer Food explain the basic principles that each individual should know all over the world to prevent foodborne diseases. Over 100 countries have reported using the "Five Keys to Safer Food."

The keys are as follows [7]:

1. keep clean;
2. separate raw and cooked;
3. cook thoroughly;
4. keep food at safe temperatures; and
5. use safe water and raw materials.

12.8   FOOD TECHNOLOGY AND SAFETY

WHO and FAO convened a meeting to publish the caveats of the use of nanotechnology in the food industry.

Nanotechnology promises enhanced working, increased flavor, dispersion and bioavailabilty to name a few. But, the implications on its impact on human health and environment has also raised several issues on safety. Nanomaterials are also finding applications to be used in animal feed and agrochemicals [8].

The recommendations may be summarized as below:

* There is a need for agreement on a specific set of clear and internationally harmonized definitions that relate to the agri-food sector, and FAO/WHO should support activities in this direction.
* There is a need to consider the whole life cycle of ENMs (engineered

nano materials) in agri-food applications. Because of potential public health implications, the use of biopersistent ENMs in the agricultural sector, which may persist or accumulate in the body or the environment, should be considered. The FAO/WHO have developed tiered approaches to risk assessment of the nanomaterials.

The WHO has also compiled a list of information on genetically modified foods in lieu of their potential in food safety.

12.9   GENETICALLY MODIFIED ORGANISMS (GMOS)

GMOs can be defined as organisms (i.e., plants, animals, or microorganisms) in which the genetic material (DNA) has been altered in a way that does not occur naturally by mating and/or natural recombination. Foods produced from or using GM organisms are often referred to as GM foods. One of the objectives for developing plants based on GM organisms is to improve crop protection.

The WHO Department of Food Safety and Zoonosis aims at assisting national authorities in the identification of foods that should be subject to risk assessment and to recommend appropriate approaches to safety assessment (Source: http://www.who.int/foodsafety/areas_work/food-technology/faq-genetically-modified-food/en/) [9].

The safety assessment of GM foods generally focuses on: (a) direct health effects (toxicity); (b) potential to provoke allergic reaction (allergenicity); (c) specific components thought to have nutritional or toxic properties; (d) the stability of the inserted gene; (e) nutritional effects associated with genetic modification; and (f) any unintended effects that could result from the gene insertion.

12.10   THE CODEX ALIMENTARIUS COMMISSION (CODEX)

Codex is the joint FAO/WHO intergovernmental body responsible for developing the standards, codes of practice, guidelines and recommendations that constitute the Codex Alimentarius, which is the international food code. Codex developed principles for the human health risk analysis of GM foods in 2003 (Source: http://www.fao.org/docrep/w9114e/W9114e04.htm) [10].

C O D E X
A L I M E N T A R I U S
International Food Standards

Codex principles do not have a binding effect on national legislation, but are referred to specifically in the Agreement on the Application of Sanitary and Phytosanitary Measures of the World Trade Organization (SPS Agreement), and WTO Members are encouraged to harmonize national standards with Codex standards. If trading partners have the same or similar mechanisms for the safety assessment of GM foods, the possibility that one product is approved in one country but rejected in another becomes smaller [10–12].

The Cartagena Protocol on Biosafety, an environmental treaty legally binding for its parties which took effect in 2003, regulates transboundary movements of living modified organisms (LMOs). GM foods are within the scope of the protocol only if they contain LMOs that are capable of transferring or replicating genetic material. The cornerstone of the protocol is a requirement that exporters seek consent from importers before the first shipment of LMOs intended for release into the environment [11–13].

Nutrigenomics studies how genetic and cellular processes relate to nutrition and health, including how people with different genetic variants respond to alternative dietary conditions and how diet can switch genes on or off. Food companies are interested in nutrigenomics because it can help them to develop and market new functional foods.

12.11 FOOD POLITICS

Food politics include the political aspects of the production, control, regulation, inspection, distribution and consumption of food. These political factors can be affected by the ethical, cultural, medical, and environmental factors. Messer et al. [15] state that protectionist trade policies, international trade agreements, famine, political instability, and development aid influence food politics [14].

In addition, climate change concerns and predictions are gaining the attention of those most concern with ensuring an adequate worldwide food supply [15].

12.12   GENETIC UNIFORMITY

Biodiversity provides the different combinations of genes, that produce the spectrum of plant varieties and animal breeds.

Varieties of different and genetically distinct varieties of major food crops owe their existence to millions of years of evolution and to careful selection and nurturing over years of agriculture. This diversity protects the crop and helps it meet the demands of different environments and human needs. Potatoes, for instance, originated in the Andes, but presently, they can be found growing across the world.

FAO estimates that a sizable portion of the genetic diversity of agricultural crops has been lost. There is increasing dependence on fewer crop varieties and, as a result, a rapidly diminishing gene pool [16]. The primary reason is that traditional species are being replaced by commercial, uniform varieties.

Industrialized agriculture favors genetic uniformity. It employs plantation of a single, high-yielding variety, a practice known as monoculture—using expensive inputs such as irrigation, fertilizer, and pesticides to maximize production. In the process, not only traditional crop varieties, but long-established farming ecosystems are depleted. Genetic uniformity invites disaster because it makes a crop vulnerable to attack—a pest or disease that strikes one plant quickly spreads throughout the crop [14].

The Irish Potato Famine of the 1840s is a notable example of the dangers of genetic uniformity [15]. A genetically uniform clone (of a single variety called Lumpers) was being cultivated. None of the few varieties of the new world potato introduced into Europe in the 1500s were resistant to a potato blight *Phytophthora infestans*) that struck Ireland in the 1840s, and caused death of the population and migration of many Irish.

In 1970, genetic uniformity left the United States maize crop vulnerable to a blight that destroyed and reduced yields by as much as 50% [16].

Over 80% of the commercial maize varieties grown in the United States at that time were susceptible to the virulent disease, southern leaf blight.

The wheat stem rust (*Puccinia graminis*) that devastated wheat fields in 1917, the elimination of all oats derived from the variety Victoria by Victoria blight (*Cochliobolus victoriae*) in the mid-1940s, and the southern corn blight (*Helminthosporium maydis*) resulted in crop loss estimated to 15% yield reduction in corn [17].

Other examples include the great Bengal famine in India in 1943 due a devastative disease (*Cochliobolus miyabeanus*) to rice [18]. An excellent example of devastation of that scale by insect pests was encountered over a century ago in France when grapevine was totally wiped out by attacks on root stocks of *Phylloxera vertifoliae* [19].

In the 1970s , grassy-stunt virus devastated rice fields, endangering the world's single most important food crop. After a four-year search that screened over 17, 000 cultivated and wild rice samples, disease resistance was found. A population of the species *Oryza nivara,* growing wild near Gonda in Uttar Pradesh was found to have a single gene for resistance to grassy-stunt virus strain 1 [20].

According to CGIAR [20], nearly 250,000–300 000 species of plants exist; 10, 000 –50, 000 are edible and only 150–200 are used as human food. The three major species, namely rice, wheat and maize, supply nearly 60% of the calorie and protein requirement to the humans. The genetic resources of plants and animals are extremely valuable to humankind. Our ability to respond to changing conditions and our capacity to maintain and enhance productivity of forest and livestock decrease as a result of loss of genetic diversity. The key to increase food security and improve the human condition lies in preserving the genetic resources.

The growing human population and rising demands for more food, and the success of such efforts like the "Green Revolution" from adoption of genetically uniform varieties in many parts of the world are the main driving force toward this narrowing of the genetic pool [20, 21].

A well-known occurrence of disease susceptibility in crops without genetic diversity is the "Gros Michel," a seedless banana variety. Due to its increased demand for this particular cultivar, growers and farmers began to use the Gros Michel banana almost exclusively. Genetically, these bananas are clones , and because of this lack of genetic diversity, they are all susceptible to a single fungus, *Fusarium oxysporum* (Panama disease); large areas of the crop were destroyed by the fungus in the 1950s [22, 23].

Tola et al. [25] write that the "Gros Michel" has been replaced by the current main banana on the market, the "Cavendish," which again is at risk of total loss to a strain of the same fungus, "Tropical Race 4" [24].

The Global Crop Diversity Trust is an independent international organization that exists to ensure the conservation and availability of crop diversity for food security worldwide. It was established through a partnership between the FAO and the Consultative Group on International Agricultural Research (CGIAR) acting through Bioversity International. The CGIAR is a consortium of International Agriculture Research Centers (IARC) and other centers that conduct research on and preserve germplasm from a particular crop or animal species. The CGIAR holds one of the world's largest off site collections of plant genetic resources in trust for the world community.

12.13 GENETIC RESOURCES AND PRESERVATION

Genetic resources are, according to the International Convention for Biodiversity, living material that includes genes of present and potential value for humans.

It has been previously discussed in the preceding sections that crop varieties with a narrow genetic pool can be completely destroyed by diseases. The plant breeders must then go back to older varieties or closely related wild species in order to find resistance genes and ensure the survival of the species.

Plant genetic resources can be conserved within or outside their natural habitats, or by combining the two alternatives [25, 26]. Outside their natural habitats, plant genetic resources are conserved in germplasm collections and genebanks. Plants are conserved according to their current or future usefulness to humans. Plant genetic resources can be conserved in their natural habitats (i.e., *in situ*), in conditions different from those of their natural habitats (i.e., *ex situ*) or in a combination of in situ and ex situ methods.

12.13.1   IN SITU CONSERVATION METHODS

In-situ conservation, the conservation of species in their natural habitats, is considered the most appropriate way of conserving biodiversity [25].

Conserving the areas where populations of species exist naturally is an underlying condition for the conservation of biodiversity. Hence, the protected areas form a central element of any national strategy to conserve biodiversity.

12.13.2   EX SITU CONSERVATION METHODS

All species can be conserved ex situ, provided they can be multiplied. Materials used include wild varieties, weedy varieties, primitive culti-vars from traditional agriculture, biotechnology products like transgenics, markers, introns etc. [26].

Germplasm may be acquired through exchange with institutions, or organizations that possess it. Upon unavailability of the same, the germ-plasm may be obtained from the region where it grows naturally. The samples should be healthy, well documented, and qualify the quarantine period in the new country.

12.13.3   GERMPLASM TRANSFER

Germplasm Transfer is achieved through the signing of an agreement between the institutes involved. The agreement stipulates the terms of both transfer and use of the material (e.g., conservation, research or production of commercial varieties). These agreements are known as "material trans-fer agreements for the exchange of genetic resources" (MTA) [27].

It is expected that the same number should be taken from every plant and in good physical and sanitary conditions. The moisture content and temperature at which the samples are to be maintained should also be con-trolled. The samples must be prevented from drying out or rotting, either of which affects their viability. If the objective is to collect seeds, harvest-ing the fruits would be advisable because this prolong the seeds' viability. The seeds can be manually extracted later. The collected seeds should be mature so that they tolerate drying without losing viability. For vegetative material, however, fresh propagules and buds should be collected so that they will reproduce later. Samples may be entire plants, tubers, rhizomes, or stakes. Plants can be collected in any container provided that it is safe and easy to transport. Plastic bags can be used for tubers, rhizomes, and stakes [25–27].

Collected samples should be kept viable until they arrive at the place of conservation. They must therefore be conditioned to prevent being damaged or contaminated. Conditioning includes cleaning or drying. Cleaning consists of removing all impurities such as stones, soil, insects, damaged or infected seeds, seeds of other species, and plant residues. Drying comprises the reducing moisture levels in the seeds to be stored. This can be carried out with silica gel, dry-air circulation equipment, or by spreading them out in thin layers, under shade, and in cool airy sites.

To prevent loss of viability of the germplasm, where possible, partial shipments of the samples should be made to the place of conservation. The shipped material should be clearly identified and be accompanied by handling instructions and the documentation stipulated by the FAO International Code.

In vitro collection consists of taking and transporting in vitro to the laboratory viable plant tissues known as explants (e.g., buds, meristems and embryos). An explant is extracted, sterilized, and cultured onto a culture medium. In vitro collection is practiced with species whose samples are difficult to handle, such as those of vegetative reproduction or unorthodox seed [25, 28]. The in vitro method has been used to collect coconut (*Cocos nucifera*), cotton (*Gossypium* spp.), cacao (*Theobroma cacao*), etc.

This is followed by primary multiplication of the germplasm material for its further propagation under optimal conditions.

The germplasm may be stored as intact plants in a field or in the form of seeds. The seeds may be dried, moisture content determined, and then stored in appropriate containers usually at low temperature and appropriate humidity levels [28].

Field genebanks maintain living plants. For plants that are perennial, arboreal, wild, semi-domesticated and heterozygous, or reproduce vegetatively, or have seeds that are short-lived or sensitive to drying, the germplasm may be stored in the field [28, 29].

There are several other techniques that are employed in storage, i.e., in vitro conditions:

In vitro genebanks are a means to overcome the disadvantages of the field genebanks and employ plant tissue culture techniques [26].

Slow growth Storage: Explants may be maintained at slow growth by low osmotic pressure of the medium, increasing osmotic pressure or plant hormones like abscisic acid. Reduction of the temperature or sources of carbon or nitrogen has also been employed. Benelli et al. [27] employed the technique for storage of selected lines of *Anthurium, Ranunculus*, and *Carex* [26].

The technique requires in between renewals at appropriate time intervals.

12.14   CRYOPRESERVATION

Cryopreservation consists of placing explants in liquid nitrogen at low temperatures (e.g., −196°C) to stop their growth while maintaining their viability and genetic and physiological stability.

Before cryopreservation, cold acclimation and preculture may be done to increase survival percentages after cryopreservation. Cold acclimation is a treatment by which plantlets are cultured at about 5°C for 1 week to 2 months. The temperature and time may depend on the germplasm.

cryoprotectants such as dimethyl sulfoxide (DMSO), ethylene glycol (EG), and glucose can be utilized as cryoprotectant [28].

There are two types of liquid-solid phase transitions in aqueous solutions.

(a) *Ice formation* is the phase transition from liquid to ice crystals, and
(b) *Vitrification* is a phase transition from a liquid to amorphous glass that avoids crystallization. Vitrification refers to the physical process by which a highly concentrated cryoprotective solution supercools to very low temperatures and finally solidifies into a metastable glass without crystallization. Vitrification had been proposed as a method for the cryopreservation of biological materials because of the potentially detrimental effects of extracellular and intracellular freezing might be avoided.

In the plant vitrification method, plant vitrification solution (PVS) is used, which is an extremely concentrated solution (7–8 M) of cryoprotectants. The most applied PVS is PVS2 solution that contains glycerol, ethylene glycol, DMSO, and sucrose in basal Murashige and Skoog medium [29].

Seed preservation at super low temperature (by vapor or liquid phase of LN) has been successfully achieved for a wide range of crop species by the standard seed bank protocol.

12.15 PLANT PATENTING

A plant patent is granted by the government to an inventor (or the inventor's heirs or assigns) who has invented or discovered and asexually reproduced a distinct and new variety of plant, other than a tuber propagated plant or a plant found in an uncultivated state. In 1930, the United States began granting patents for plants and in 1931, the first plant patent was issued to Henry Bosenberg for his climbing, ever-blooming rose.

Under patent law, the inventor of a plant is the person who first appreciates the distinctive qualities of a plant and reproduces it asexually. In other words, a plant can be created (i.e., by breeding or grafting) or it can be "discovered." Plants discovered in "the wild" or uncultivated state cannot be patented because they occur freely in nature.

Countries who are members of the World Trade Organization (WTO) are obliged by Article 27.3(b) of the WTO Agreement on Trade Related Aspects of Intellectual Property Rights (TRIPS) to "provide for the protection of plant varieties either by patents or by an effective *sui generis* system or by any combination thereof" [30].

Plant varieties can be protected in the USA under a system of plant patents or under a system of utility patents or under the Plant Variety Protection Act (PVPA). The Plant Patent Act (35 U.S.C. §§ 161–164 (1994)) [31] makes available patent protection to new varieties of asexually reproduced plants. Under this scheme, a plant variety must be novel and distinct and the invention, discovery, or reproduction of the plant variety must not be obvious. One of the disadvantages of the scheme is that only one claim, covering the plant variety, is permitted in each application [30, 31].

This protection is limited to a plant in its ordinary meaning:

- A living plant organism that expresses a set of characteristics determined by its single, genetic makeup or genotype, which can be duplicated through asexual reproduction, but which cannot otherwise be "made" or "manufactured."
- Sports, mutants, hybrids, and transformed plants are comprehended; sports or mutants may be spontaneous or induced. Hybrids may be natural, from a planned breeding program, or somatic in source. While natural plant mutants might have naturally occurred, they must have been discovered in a cultivated area.
- Algae and macro fungi are regarded as plants, but bacteria are not.

Patents to plants that are stable and reproduced by asexual reproduction, and not a potato or other edible tuber reproduced plant, are provided for by Title 35 United States Code, Section 161 which states [31]:

"Whoever invents or discovers and asexually reproduces any distinct and new variety of plant, including cultivated sports, mutants, hybrids, and newly found seedlings, other than a tuber propagated plant or a plant found in an uncultivated state, may obtain a patent therefor, subject to the conditions and requirements of title. "(Amended September 3, 1954, 68 Stat. 1190).

The other stipulations for patentability are:

- That the plant was invented or discovered and, if discovered, that the discovery was made in a cultivated area.
- That the plant is not a plant that is excluded by statute, where the part of the plant used for asexual reproduction is not a tuber food part, as with potato or Jerusalem artichoke.
- That the person or persons filing the application are those who actually invented the claimed plant; i.e., discovered or developed and identified or isolated the plant, and asexually reproduced the plant.
- That the plant has not been sold or released in the United States of America more than one year prior to the date of the application.
- That the plant has not been enabled to the public, i.e., by description in a printed publication in this country more than one year before the application for patent with an offer to sale; or by release or sale of the plant more than one year prior to application for patent.

- That the plant be shown to differ from known, related plants by at least one distinguishing characteristic, which is more than a difference caused by growing conditions or fertility levels, etc.
- The invention would not have been obvious to one skilled in the art at the time of invention by applicant.

In Europe, European Patent Convention (EPC) takes account of International Convention for the Protection of New Varieties of Plants (UPOV) and, in Article 53(b) [32], specifically excludes the patenting of "plant or animal varieties or essentially biological processes for the production of plants or animals," explaining that "this provision shall not apply to microbiological processes or the products thereof." Rule 23b(5) of the EPC explains that a process for the production of plants and animals is essentially biological "if it consists entirely of natural phenomena such as crossing or selection." This language is replicated in the EU Biotechnology Directive, which in Article 4.1 excludes from patentability [33, 34]:

(a) plant and animal varieties; and
(b) essentially biological processes for the production of plants or animals. Article 2.2 states that a process for the production of plants or animals is essentially biological "if it consists entirely of natural phenomena such as crossing or selection."

The patenting of inventions in Australia is governed by the Patents Act 1990. In Australia, two types of patent are available: standard and innovation patents [35]. The law on patentable subject matter in Australia is, and always has been, among the most liberal of any patent system. Along with Europe, Japan, and the United States, Australia is one of only three countries in which patents may be obtained for new plant (and animal) varieties (Plant Law Practice, 2011) [35].

12.16   ADVANTAGES AND DISADVANTAGES OF BIOTECHNOLOGY

Biotechnology has ushered in shifts in agriculture through the use of transgenic crops that have enhanced yield, enhanced nutrients or express resistance to pests.

Monsanto states "After 13 years of use on more than 2 billion acres (800 million hectares) worldwide, plant biotechnology delivers proven economic and environmental benefits, a solid record of safe use and promising products for our future" [36, 37].

In 1994, the FLAVR SAVR tomato, modified to delay premature fruit softening made its appearance [36–38]. In 2014, genetically modified (GM) crops were grown by 18 million farmers in 28 countries on a total surface of 181.5 million hectares, which correspond to already 13% of the world's arable surface [36–38]. Globally, 82% of the total crop area for soybeans, 68% for cotton, 30% for maize and 25% for oilseed rape were planted with GM varieties in 2014 [39].

The use of pesticides may have reduced with the development of various insect/pest- resistant crops.

Plant breeding has been augmented by the development of seed-chipping devices (The seed chipping innovation was carried out by Kevin Deppermann, at Monsanto [37]. It employs the use of a computer- controlled seed chipper that allows seed breeders to know the exact DNA makeup of a seed before it is planted. By knowing the DNA before planting, those seeds that do not have the desired genetic makeup are never planted but discarded).

DNA fingerprinting methods have also aided the screening of desired allelic combinations for plant breeding technologies.

The regulations on biotechnology over nations sometimes stifle such research as there is no clear boundary in the rules involved in the biotechnological legislations [40].

Genome editing shows great promise that employs the use of endonucleases to edit genomes by insertion, substitution, or deletion, predetermined sequences in genomes. Zinc-finger nucleases (ZFN), transcription activator-like effector nucleases (TALENs), and clustered regularly interspaced short palindromic repeats (CRISPRs) are a few endonucleases that are used to program such effects [41, 42].

Oligonucleotide-directed mutagenesis (ODM) uses short oligonucleotides synthesized that are similar to the DNA sequence but contain a change to be introduced that is incorporated in subsequent repair [41].

Various gene knockouts can be generated using NHEJ (Nonhomologous End Joining) used for removing harmful variations or antinutrients or introduction of resistance to pests by altering specific recognition pathways [42].

Wang et al. [44] produced broad-spectrum mildew-resistance in wheat by knocking out a genetic locus suspected of suppressing plant defense against mildew [43]. Plants where all alleles were successfully knocked out by TALENs or CRISPR were highly resistant to mildew.

The first commercial application of genome editing was developed by Cibus Global. In March 2014, they received regulatory approval from Canadian Food Inspection Agency and Health Canada to commercialize a novel sulfonylurea-tolerant Canola generated using their proprietary Genome Repair Oligonucleotide technology (Canadian Food Inspection Agency, 2013, AgCanada News, 2014) [44, 45].

The use of RNA- based mechanisms has been employed for various purposes. For example, two soy varieties (Monsanto's Vistive Gold and DuPont-Pioneer's Plenish™) possess high oleic/low linolenic oil giving better heat stability for frying, longer fry life, and improved flavor of fried products produced partly through a gene silencing effect [46].

Arctic Apples (Okanagan Specialty Fruits) that received regulatory approval from APHIS in March 2015 (USFDA, 2014) have less of the enzymes polyphenol oxidase that cause browning.

Transgenic, papaya ringspot-virus (PRSV)- resistant papaya trees were introduced in Hawaii in 1998 after the papaya production was on the verge of collapse because of a devastating outbreak of PRSV infections [47]. They were rapidly taken up by the large majority of the papaya farmers in almost 90% on the papaya cultivation surface in Hawaii, and are credited with saving the Hawaii papaya industry from extinction [47].

A positive side-effect of the switch to biotech plant varieties has been a pronounced reduction in insecticide quantities used on insect-resistant Bt-crops (−41.7%), and the possibility to switch to more environmentally benign herbicides with herbicide- tolerant crops [48].

In 2014 about 95% of the cotton area was planted with transgenic varieties. Bt cotton has reduced the dependency on chemical pest control, increased yields and profits for smallholder farmers in a sustainable way over a long period, and has thereby contributed to a positive economic and social development in India. Increased farmer's income translates also into increased food security for cotton farmers in India [49].

Facing the challenge of a world population that is expected to reach an estimated 9 billion in 2050, food production globally has to increase by about 70 % in order to feed the world. The difficulty in managing this daunting task is that we have to produce more food with less arable land. Moreover, climate change adds yet another challenge to food security (FAO High Level Expert Forum; Rome 12–13 October 2009, http://www.fao.org) [50].

There is need for advancement in biotechnology and crops to feed the hungry. Few examples are China, India, Brazil, Argentina, and South Africa. They grow about 46 % of global GM crops, and have ~ 40 % of the world population (see detailed information provided by www.isaaa.org) [51].

Golden Rice (GR), a GM rice rich in β-carotene for use as a source of vitamin A, has been reported to overcome dietary lack of nutrients in 2000. Rice produces β-carotene in the leaves but not in the grain, where the biosynthetic pathway is turned off during plant development. In GR , two genes have been inserted into the rice genome by genetic engineering to restart the carotenoid biosynthetic pathway, leading to the production and accumulation of β-carotene in the grains [52].

The consumption of anthocyanin-rich food promotes health. Many recent studies of anthocyanin-rich fruits such as blueberry, bilberry, and cranberry support claims that anthocyanin consumption promotes health. Purple GM tomatoes with enhanced anthocyanin accumulation have been produced and animal studies have proven its health value [53, 54].

Plant biotechnology has served as an effective instrument for genetic engineering of plants for cost effective production of medicines. A plant-produced veterinary vaccine against Newcastle disease in poultry was approved in 2006 by US Department of Agriculture Center for Veterinary Biologics (source: www.thepoultrysite.com) [55].

12.17 PLANT-BASED PRODUCTION OF THERAPEUTICS

The advantages of plant-based expression systems include high scalability, low upstream costs, biocontainment, lack of human or animal pathogens, and ability to produce target proteins with desired structures and biological functions [56, 57]. Using transgenic and transient expression in whole plants or plant cell culture, a variety of recombinant subunit vaccine candidates, therapeutic proteins, including monoclonal antibodies, and dietary proteins have been produced. Some of these products have been tested in early phase clinical trials and show safety and efficacy. Among those are mucosal vaccines for diarrheal diseases, hepatitis B and rabies; injectable vaccines for non-Hodgkin's lymphoma, H1N1 and H5N1 strains of influenza A virus, and Newcastle disease in poultry; and topical antibodies for the treatment of dental caries and HIV.

The shared diseases between animals and humans are known as zoonotic diseases and spread infectious diseases among humans. About 75% of emerging infectious diseases in humans have been reported to originate from zoonotic pathogens. Another concern is the development of antibiotic resistance. For the control of the diseases, live/attenuated vaccines have been used that possess challenges like high cost and the daunting task of individual injection of large number of animals/birds.

Porcilis-PCV2 and Suvaxyn PCV2 for pigs, Periovac for dogs, AquaVac ERM, AquaVac Furuvac, and AquaVac Vibrio for fish are a few commercialized and licensed vaccines against veterinary diseases. PreveNile against horses, Vaxxitek HVT+IBD against poultry, Bovilis IBR Marker against cattle, RECOMBITEK Canine Parvo against dogs, RECOMBITEK Corona MLV against dogs, and Enterisol Ileitis against pigs are commercially available live virus vaccines.

The United States Department of Agriculture (USDA) approved the world's first plant-based vaccine in 2006. Dow AgroSciences received approval for the first plant-based vaccine against Newcastle disease virus (NDV) from the USDA. Dow Agro Sciences used tobacco suspension cell lines to develop a plant-based vaccine (injectable) against NDV. Although this vaccine was approved by the USDA in 2006 after showing 90% protection against a challenge with NDV virus, it was not commercialized.

Edible vaccines are actually recombinant vaccines in which selected antigens against a particular pathogen are introduced into a plant. There are several advantages of using plants for producing therapeutics like cost effectiveness, safety, easy delivery against needles, absence of toxins, and storage for some time [58]. Wen et al. [60] report the development of wheat with reduction in immunogenic prolamines that could help in the development of gluten free wheat [59].

Despite the mounting evidence showing the importance of GM plants in food, feed, health industries, and bioeconomy, GM plants experience strong opposition worldwide, more than any other modern technology has encountered. GM crops and products in particular are met by regulatory constraints and must cope with a long process for approval in many countries, especially in Europe.

Few potential risks have been raised with respect to the use of biotechnology in agriculture. An "allergy" is a hypersensitive immune response that occurs when a person comes into contact with specific substances called allergens. Ninety percent of food allergies are caused by the common allergens in peanuts, tree nuts, milk, eggs, wheat, soy, shellfish, and fish.

Potential limitations with the use of RNAi mediated methods have been raised. For instance, post- transcriptional silencing of genes in plants may be mobile, causing its effect to spread to the plants and thus raising a hitch in tissue- specific silencing. siRNAs comprise a population of molecules representing the entire sequence of the dsRNA trigger. Although this sequence heterogeneity could make it easy to silence a family of related genes with only one construct, it also opens the door to off-target effects, in which genes with regions of homology to the intended target get silenced unintentionally.

A new safety concern that is unique to RNAi approaches has recently been raised by a report that miRNAs made in plants (MIR168a is abundant in rice) are taken up by humans and other mammals when they eat plants and bind human/mouse low-density lipoprotein receptor adapter protein 1 (LDLRAP1) mRNA [61]. Though the finding has been challenged, we can say that it is essential to check for the effects or implications of such potential "non-target" binding.

The perception of new food technologies is strongly affected by a complex and deeply rooted set of personal values and attitudes. A very important factor seems to be the personal importance of the naturalness of a product. Additional factors affecting personal attitudes about GM crops and food are cultural values like an aversion to be invaded or dominated by foreign food culture, and conflicts with the religious or moral belief system, or the perceived natural order of things [62–64]. A few other important factors include the attitude of the consumers across various countries as each country has different perceptions. Emphasis is also placed on the political framework of the country.

12.18 SUMMARY

Food ethics raises issues and asks questions in relation to food all along the value chains. It also puts things into relation with each other. Key global food safety concerns include the spread of microbiological hazards (including bacteria such as *Salmonella* or *Escherichia coli*, etc.); chemical food contaminants; assessments of new food technologies (such as genetically modified food); and strong food safety systems in most countries to ensure a safe global food chain. WHO recognizes that food safety is essential given today's global scenario and that consumers must be aware of food safety. The value added to an organization with a certified food safety system lies in the efforts made by the organization to maintain that system and its commitment to continuously improve its performance. Over 100 countries have reported using the "Five Keys to Safer Food." Food politics include the political aspects of the production, control, regulation, inspection, distribution, and food consumption.

12.19 REVIEW QUESTIONS

1. Briefly describe the role of
    a.   Global Food Safety Initiative.
    b.   British Retail Consortium
    c.   The International Association for Food Protection
    d.   Codex Alimentarius Commission
2. Mention the five keys to safer food program.
3. Describe the methods of preservation of genetic resources.
4. Write a note on plant patenting.
5. How have plants been used for the production of therapeutics?

KEYWORDS

- FAO
- GFSI
- GMO
- preservation
- food safety
- germplasm
- IAFP

REFERENCES

1. Food and Agriculture Organization (FAO), (1996). Rome Declaration on World Food Security and World Food Summit Plan of Action. *World Food Summit*, Rome.
2. http://www.mygfsi.com/about-us/about-gfsi/what-is-gfsi.html.
3. https://www.cert-id.com/Certification-Programs/BRC-Certification.aspx.
4. https://www.foodprotection.org/.
5. http://www.iufost.org/about-iufost.
6. WHO estimates of the global burden of foodborne diseases, eds: World Health Organization, Dec 2015. http://www.who.int/foodsafety/publications/foodborne_disease/fergreport/en/.
7. http://who.int/foodsafety/areas_work/food-hygiene/5keys/en/.
8. FAO/WHO Expert meeting on the application of nanotechnologies in the food and agriculture sectors: potential food safety implications Meeting report. March 2012. http://www.fao.org/docrep/012/i1434e/i1434e00.pdf.
9. http://www.who.int/foodsafety/areas_work/food-technology/faq-genetically-modified-food/en/.
10. Jessica Fanzo, (2015). Ethical issues for human nutrition in the context of global food security and sustainable development. *Global Food Security*. vol. 7, pp. 15–23.
11. Matthias Kaiser, (2016). Food ethics: a wide field in need of dialogue. *Food Ethics*, *1*(1), pp. 1–7.
12. Anne Algers, (2016). *Food Ethics*, *1*(1), pp. 1–7.
13. http://www.fao.org/.
14. FAO/WHO Expert meeting on the application of nanotechnologies in the food and agriculture sectors: potential food safety implications. Meeting report.
15. *Messer, E., & Cohen, M., (2007)." Conflict, food insecurity and globalization." Food, culture and society: An International Journal of Multidisciplinary Research, 2(10), 297–315. doi: 10.2752/155280107x211458.*
16. Report on the State of the World's Plant Genetic Resources for Food and Agriculture. International Technical Conference on Plant Genetic Resources Leipzig, Germany, (1996). ftp://ftp.fao.org/docrep/fao/meeting/015/aj633e.pdf.
17. Mubeen, S., et al., (2017). Study of southern corn leaf blight (SCLB) on maize genotypes and its effect on yield. *Journal of the Saudi Society of Agricultural Sciences*, http://dx.doi.org/10.1016/j.jssas.06.006, *16*(3), pp. 210–217.
18. *Perry, M., Rosenzweig, C., & Livermore, M., (2005). "Climate change, global food supply and risk of hunger." Philosophical transactions of the royal society, Biological Sciences, 1463(350), 2125–2138.*
19. Miguel, A., Altieri & Clara I. Nicholls & Alejandro Henao & Marcos A. Lana. Agroecology and the design of climate change-resilient farming systems, Agron. Sustain. Dev. DOI 10.1007/s13593–015–0285–2.
20. Gemechu, K., Endashaw, B., Muhammad, I., & Kifle, D., (2012). Genetic vulnerability of modern crop cultivars: Causes, mechanism and remedies. *International Journal of Plant Research*, 2(3), 69–79, DOI: 10.5923/j.plant.20120203.05.
21. http://www.fao.org/docrep/004/v1430e/V1430E04.htm.

22. Gemechu, K., Endashaw, B., Muhammad, I., & Kifle, D., (2012). Genetic vulnerability of modern crop cultivars: causes, mechanism and remedies. *International Journal of Plant Research* p-ISSN: 2163–2596 e-ISSN: 2163–260X., *2*(3), 69–79.

23. Rubenstein, K. D., Heisey, P., Shoemaker, R., Sullivan, J., &Frisvold, G. (2011). Crop genetic resources: An economic appraisal. In Plant Genetic Resources and Food Security (pp. 37-88). Nova Science Publishers, Inc.

24. Randy, C. P., (2005). *Panama Disease*: An old nemesis rears its ugly head part 1: The beginnings of the banana export trades." www.apsnet.org.

25. Tola, E., (2015). "Banana variety risks wipeout from deadly fungus wilt." *The Guardian.* https://www.theguardian.com/global-development/2015/jan/21/banana-deadly-fungus-wilt-fao-cavendish-fusarium.

26. Sildana, J., & Margarita, B., (2002). Ex Situ conservation of plant genetic resources training module. *Biodiversity International.* International Plant Genetic Resources Institute, Cali, Colombia. https://www.bioversityinternational.org/uploads/tx_news/Ex_Situ_conservation_of_plant_genetic_resources_1252.pdf.

27. Benelli, C., Ozudogru, E. A., Lambardi, M., & Dradi, G., (2012). In vitro conservation of ornamental plants by slow growth storage. *Acta Hort. (ISHS), 961*, 89–93.

28. John, H. B., Wolfang, S., & Siebeck, E., (1994). Material transfer agreements in genetic resources exchange – the case of the International Agricultural Research Centers. *Issues in Genetic Resources No. 1.* International Plant Genetic Resources Institute, Rome, Italy.

29. Cryopreservation of Plant Genetic Resources Daisuke Kami National Agricultural Research Center for Hokkaido Region Japan, (2015). Published by in tech., *Breeding Science, 65*(1), pp. 41–52.

30. Takao, N., & Miriam, V. A., (2015). Cryopreservation for preservation of potato genetic resources. *Breeding Science, 65*(1), pp. 41–52.

31. Overview: the TRIPS Agreement. January (1995). https://www.wto.org/english/tratop_e/trips_e/intel2_e.htm.

32. General Information About 35 U. S. C. 161 Plant Patents.

33. European Patent Convention (EPC 1973). http://www.epo.org/law-practice/legal-texts/html/epc/1973/e/ar53.html.

34. http://www.nolo.com/legal-encyclopedia/plant-patents.html.

35. Michael, B., (2012). Patenting of plant varieties and plant breeding methods.. *J. Exp. Bot., 63*(3), 1069–1074.

36. Plant patent law and practice: Australia (2011), North America and Europe Discussion Paper ARC Discovery Project DP0987639, Promoting plant innovation in Australia: Maximizing the benefits of intellectual property for Australian Agriculture. Australian Centre for Intellectual Property in Agriculture.

37. Greentumble Editorial Team, (2017). Advantages and Disadvantages of Using Biotechnology in Agriculture https://greentumble.com/advantages-and-disadvantages-of-using-biotechnology-in-agriculture/.

38. http://www.monsanto.com/products/pages/benefits-of-plant-biotechnology.aspx.

39. https://croplife.org/wp-content/uploads/pdf_files/Benefits-of-Plant-Biotechnology.pdfv.

40. JISAAA, (2016). Global Status of Commercialized Biotech/GM Crops: 2016. ISAAA Brief No. 52. ISAAA: Ithaca, NY. Available on: http://www.haseloff-lab.org/resources/Part2SynBio_refs/Lecture-1/isaaa-brief-52-2016.pdf.

41. Benefits of genetically modified crops for the poor: household income, nutrition, and health. New biotechnology. Transgenic plants for food security in the context of development – *Proceedings of a Study Week of the Pontifical Academy of Sciences, 27*(5), 2010, pp. 552–557.

42. Sauer, N. J., Narváez-Vásquez, J., Mozoruk, J., Miller, R. B., Warburg, Z. J., Woodward, M. J., & Gocal, G. F. W., (2016). Oligonucleotide-mediated genome editing provides precision and function to engineered nucleases and antibiotics in plants. *Plant Physiology, 170*(4), 1917–1928. http://doi.org/10.1104/pp.15.01696.

43. Voytas, D. F., & Gao, C., (2014). Precision genome engineering and agriculture: Opportunities and regulatory challenges. *PLoS Biol., 12*(6), e1001877, doi:10.1371/journal.pbio.1001877.

44. Wang, Y., Cheng, X., Shan, Q., Zhang, Y., Liu, J., Gao, C., et al., (2014). Simultaneous editing of three homoeoalleles in hexaploid bread wheat confers heritable resistance to powdery mildew. *Nat. Biotechnol., 32*, 947–951.

45. Canadian Food Inspection Agency (2013), *AgCanada News* (2014).

46. Gonsalves, D., & Ferreira, S. (2003). Transgenic papaya: A case for managing risks of papaya ringspot virus in Hawaii. Transgenic papaya: A case for managing risks of papaya ringspot virusin Hawaii. *Plant Health Progress,* DOI 10.1094/PHP-2003–1113–03-RV.

47. Jones, H. D., (2015). Challenging regulations: Managing risks in crop biotechnology. *Food and Energy Security, 4*(2), 87–91, doi:10.1002/fes3.60.

48. Gonsalves, C. V., & Gonsalves, D., (2014). The Hawaii papaya story. In: Smyth, S. J., Phillips, P. W. B., Castle, D., editors. *Handbook on Agriculture, Biotechnology and Development*. Edward Elgar Publishers, Northhampton, MA, USA,. pp. 642–660.

49. Klümper, W., & Qaim, M., (2014). A meta-analysis of the impacts of genetically modified crops. *PLoS One., 9*(11), e111629.

50. Kathage, J., & Qaim, M., (2012). Economic impacts and impact dynamics of Bt (Bacillus thuringiensis) cotton in India. *Proc Natl Acad Sci USA, 17, 109*(29), 11652–11656.

51. FAO, (2009) high level expert forum, Rome, http://www.fao.org.

52. www.isaaa.org.

53. www.goldenrice.org.

54. Martin, C., Butelli, E., Petroni, K., & Tonelli, C., (2011). How can research on plants contribute to promoting human health? *Plant Cell., 23*, 1685–1699.

55. Glover, B. J., & Martin, C., (2012). Anthocyanins. *Curr. Biol., 22*, 147–150. doi:10.1016/j.cub.2012.01.021.

56. www.thepoultrysite.com.

57. Gary, W., (2014). Biopharmaceutical benchmarks. *Nature Biotechnology, 32*, 992–1000, doi:10.1038/nbt.3040.

58. Yusibov, V., Streatfield, S. J., & Kushnir, N., (2011). Clinical development of plant-produced recombinant pharmaceuticals: vaccines, antibodies and beyond. *Hum. Vaccin., 7*(3), 313–21.

59. Naila, S., & Henry, D., (2016). Plant-based oral vaccines against zoonotic and non-zoonotic diseases. *Plant Biotechnol. J.*, *14*(11), 2079–2099.

60. Wen, S., et al., (2012). Structural genes of wheat and barley 5-methylcytosine DNA glycosylases and their potential applications for human health. *PNAS, 109*, 20543–20548.

61. Zhang, L., Hou, D., Chen, X., Li, D., Zhu, L., Zhang, Y., Li, J., Bian, Z., Liang, X., Cai, X., Yin, Y., Wang, C., Zhang, T., Zhu, D., Zhang, D., Xu, J., Chen, Q., Ba, Y., Liu, J., Wang, Q., Chen, J., Wang, J., Wang, M., Zhang, Q., Zhang, J., Zen, K., Zhang, C. Y., (2012). Exogenous plant MIR168a specifically targets mammalian LDLRAP1: evidence of cross-kingdom regulation by microRNA. *Cell. Res., 22*(1), 107–126.

62. Jihong, L. C., & Peng, Z., (2013). Plant biotechnology for food security and bio-economy. *Plant Molecular Biotechnology*, *83*(1), pp. 1–3.

63. Huw, D. J., (2015). Challenging regulations: Managing risks in crop biotechnology. *Food Energy Secur.*, *4*(2), 87–91.

64. Ania, W., (2003). Use of biotechnology in agriculture – Benefits and risks. *CTAHR – Biotechnology,* (revised) BIO-3. https://www.ctahr.hawaii.edu/oc/freepubs/pdf/bio-3.pdf.

65. Public Acceptance of Plant Biotechnology and GM Crops, (2015). *Jan. M. Lucht. Viruses*, *7*(8), 4254–4281.

INDEX

T - #0822 - 101024 - C364 - 229/152/16 - PB - 9781774631683 - Gloss Lamination